STRUCTURAL DETAILS FOR MASONRY CONSTRUCTION

Morton Newman, P.E.

PROJECT EDITOR: Jeremy Robinson

McGRAW-HILL BOOK COMPANY

New York St. Louis San Francisco Auckland
Bogotá Hamburg London Madrid Mexico
Milan Montreal New Delhi Panama
Paris São Paulo Singapore
Sydney Tokyo Toronto

Library of Congress Cataloging-in-Publication Data

Newman, Morton.
Structural details for masonry construction.

"The material in this volume has been published previously in Standard structural details for building construction "—T.p. verso.
Includes index.
1. Masonry. 2. Building—Details—Drawing.
I. Newman, Morton. Standard structural details for building construction. II. Title.
TH1199.N48 1988 693'.1 87-20403
ISBN 0-07-046361-1 (pbk.)

1234567890 KGP/KGP 893210987

ISBN 0-07-046361-1

Printed and bound by Arcata Graphics/Kingsport.

The material in this volume has been published previously in *Standard Structural Details for Building Construction* by Morton Newman.

To my mother

Contents

About This Book

Since publication in 1968, over 30,000 copies of the hardcover edition of Morton Newman's STANDARD STRUCTURAL DETAILS FOR BUILDING CONSTRUCTION have been sold to architects, engineers, drafters, and others concerned with the design of building structure and the communication of that design to those responsible for accomplishing it in construction.

Now the publishers have made it possible for persons interested in one particular type of structure—wood, concrete, masonry, or steel—to purchase just the section or sections of *Standard Structural Details* which they need. The hardcover book has been split up into four separate low-cost softcover editions:

Structural Details For Wood Construction
Structural Details For Concrete Construction
Structural Details for Masonry Construction
Structural Details for Steel Construction

In this book, the designer will find a host of proven designs in time-tested details for concrete block walls; concrete block columns and pilasters; concrete block walls supporting floors, roofs, steel framing, and concrete slabs; brick walls, pilasters, and columns; brick walls and steel columns; brick walls supporting wood floors, roofs, and steel framing; typical brick wall sections; and concrete block retaining walls.

Introductory material has been revised to reflect code changes since the original edition, and each page of details appears on the right, while facing pages incorporate a preprinted grid for drawings, notes, and ideas the reader may wish to preserve.

Preface

The purpose of this series of books is to provide a graphic means of communication between architects, engineers, contractors, and students who are engaged in the design and construction of buildings. The four basic structural materials that are employed in building construction are wood, concrete, masonry, and steel. In the application of these materials many standard details and methods of construction have been developed. For several years the author found it quite useful to collect and index standard structural details for the preparation of structural drawings of buildings. The use of structural graphic standards reduced the cost of production of structural drawings and also helped to facilitate the communication of information between all of the personnel who were involved in the design and construction of a building. No claim is made for the originality of the details in these books as they are standard methods of construction and they are extensively used throughout the construction industry.

These books consist of a series of drawings of standard structural details that are most frequently employed in building construction. The details are presented individually and in their most basic and general form. A brief description is given for each detail pertaining to the material used, the type of condition shown, and its method of construction. In no instance should a book be considered or used as a substitute for the engineer or as a shortcut method of engineering. It is the function of the engineer to verify the use of any detail and to determine the sizes, dimensions, and all other pertinent information that will be essential to its use in a particular building design. The details are separated and arranged into four books with respect to the type of construction materials used: wood, concrete, masonry, or steel. In some instances two types of construction materials are used in the same detail. The author endeavored to place each detail in the related book and in the sequence of its use in building construction so that it could be readily located. Also, the index for this book has been set up so that any particular detail that may be sought can be easily located.

The engineering information presented in these books is in accordance with the basic requirements of The American Institute of Steel Construction, The American Concrete Institute, The International Conference of Building Officials Uniform Building Code, "The West Coast Lumbermen's Douglas Fir Use Book," and The Concrete Masonry Association of California. Standard details and construction methods evolve from the structural design requirements. Many excellent books on structural design and analysis are available to the practicing engineer and student; there is also a great need for applied practical information related to structural drafting and the use of construction materials. Expanding technology in the fields of building engineering and construction has created a situation which demands that the structural drawings be more complete and therefore more complex.

The purpose of structural drawings is to communicate the engineer's design requirements to the various contractors and material fabricators.

To achieve total communication, the structural drawings should be clear and complete, the general presentation of information should be in a logical sequence, all sections and details should be shown and clearly referenced, and any field conditions should be considered on the drawings. A good set of structural drawings will ensure that the building is constructed in accordance with the engineering design requirements and that construction delays and unnecessary additional costs are avoided.

The engineer's work is the prime factor in the successful design and construction of a building; however, in the final event, his or her work is directly dependent upon the intelligence and integrity of the workers on the construction job, particularly at the supervisory level. Poor fieldwork in terms of accuracy and material quality control will negate a great deal of engineering effort. Building construction requires a high degree of teamwork between the engineers and the contractors. Each party should have a working knowledge of the other's functions and responsibilities. The author hopes that these books will serve as communication tools that will improve the quality of engineering and construction. Also, engineering and architectural students can use this book as a source of information to familiarize themselves with the methods and materials of construction. As students use the information presented in these books, they will increase their ability to translate structural engineering calculations into practical applications.

I would like to acknowledge the very able assistance of Bruce L. Ward, who drew the details shown on the following pages and assisted in assembling the information into book form. Also, I would like to thank Jack Clark for his advice and encouragement, and acknowledge the assistance of Bogdan Todorovic in the early stages of these books.

Morton Newman
Civil Engineer

Abbreviations

Term	Abbreviation
Adjustable	Adjust.
Alternate	Alt.
American Concrete Institute	A.C.I.
American Institute of Steel Construction	A.I.S.C.
American Society of Testing and Materials	A.S.T.M.
Architect	Arch.
Area	A.
Beam	Bm.
Block	Blk.
Blocking	Blkg.
Bottom	Bott.
Building	Bldg.
Calculations	Calcs.
Ceiling	Ceil.
Cement	Cem.
Center Line	C.L.
Channel Stud	C.S.
Civil Engineer	C.E.
Clear	Clr.
Column	Col.
Concrete	Conc.
Connection	Conn.
Construction	Constr.
Continuous	Cont.
Cubic	Cu.
Deflection	Defl.
Depression	Depr.
Detail	Det.
Diagonal	Diag.
Diameter	Dia.
Dimension	Dim.
Discontinuous	Disc.
Double	Dbl.
Drawing	Drwg.
Each	Ea.
Elevation	El. or Elev.
Engineer	Engr.
Equal	Eq.
Equipment	Equip.
Existing	Exist.
Expand	Exp.
Expose	Expo.
Exterior	Ext.
Fillet	Fill.
Finish	Fin.
Floor	Flr.
Foot	Ft.
Footing	Ftg.
Foundation	Fdn.
Framing	Frmg.
Gauge	Ga.
Glued Laminated	Gl. Lam
Grade	Gr.
Grout	Grt.
Gypsum	Gyp.
Hanger	Hngr.
Height	Ht.
Hook	Hk.
Horizontal	Horiz.
Inch	In.
Inclusive	Incl.
Inside Diameter	I.D.
Interior	Int.
Joint	Jnt.
Joist	Jst.
Lag Screw	L.S.
Laminated	Lam.
Lateral	Lat.
Light Weight	Lt. Wt.
Machine	Mach.
Masonry	Mas.
Maximum	Max.
Membrane	Memb.

Metal	Met. or Mtl.
Minimum	Min.
Moment of Inertia	I
Nails	d (penny)
Natural	Nat.
Number	No. or #
On Center	O.C.
Opening	Opng.
Opposite	Opp.
Outside Diameter	O.D.
Panels	Pnls.
Partition	Part.
Penetration	Pen.
Plaster	Plas.
Plate	Pl.
Plywood	Plywd.
Pounds per Cubic Foot	P.C.F.
Pounds per Square Foot	P.S.F.
Pounds per Square Inch	P.S.I.
Pressure	Press.
Radius	R.
Rafter	Rftr.
Rectangular	Rect.
Reinforcing	Reinf.
Required	Reqd.
Riser	R.
Roof	Rf.
Room	Rm.
Round	ϕ
Schedule	Sched.
Section	Sect.
Section Modulus	S.
Seismic	Seis.
Sheathing	Shtg.
Sheet	Sht.
Spacing	Spcg.
Specification	Spec.
Spiral	Sp.
Stagger	Stgr.
Standard	Std.
Steel	Stl.
Steel Joist	S.J.
Stiffener	Stiff.
Stirrup	Stirr.
Structural	Struct.
Structural Steel Tube	S.S.T.
Square	Sq.
Symmetrical	Sym.
Thick	Thk.
Through	Thru.
Tread	Tr.
Ultimate	Ult.
Ultimate Stress Design	U.S.D.
Uniform Building Code	U.B.C.
Utility	Util.
Vertical	Vert.
Volume	Vol.
Waterproof	W.P.
Weight	Wt.
Welded Wire Fabric	W.W.F.
Wide Flange	W.F.
With	W/
Working Stress Design	W.S.D.

INTRODUCTION

The details in this volume are divided into three different general categories. The first part consists of drawings of hollow-unit masonry or concrete block construction. These details are arranged in the following sequence: wall intersections, columns and pilasters, wall opening lintels and jambs, wall sections with wood floors and roofs, and wall sections and concrete slabs. The second part consists of drawings of brick details arranged in the following sequence: columns and pilasters, beam and lintel sections, typical masonry wall sections and reinforcing steel placement, wall sections and wood floors and roofs, wall sections and concrete slabs, and wall sections and steel-beam connections. The third section is a series of drawings of concrete block retaining walls which are designed to support various grades and slopes.

Hollow masonry units are generally known as concrete blocks. The material used for the manufacture of this type of masonry unit consists of sand, cement, and natural crushed rock. Light-weight hollow masonry units use aggregates composed of coal cinders, slag, shale, or volcanic ash, depending on the materials available from natural sources where the units are manufactured. The sizes and shapes of the many standard manufactured hollow masonry units are shown in the drawings at the beginning of the detail drawings. These sizes and shapes are in accordance with the requirements of the Concrete Masonry Association of California.

Table 1 gives the A.S.T.M. designation numbers for the various types of concrete blocks.

Concrete hollow unit masonry is used to construct vertical structural members such as walls, beams, pilasters, columns, and foundation walls. The blocks are stacked vertically and joined to the adjacent block's horizontal and vertical surfaces by a mortar joint. Since the actual dimensions of the units are $\frac{3}{8}''$ less than the nominal dimensions, the mortar joints are made so that the finished construction will equal an even inch. Except for $\frac{3}{4}$ length units, the nominal dimensions of the lengths of concrete hollow masonry units are in mul-

Table 1.
A.S.T.M. Specifications for Concrete Block Masonry

Masonry unit	Grade of concrete	A.S.T.M. Spec. No.
Hollow load bearing	N,S	C 90
Solid load bearing	N,S	C 145
Hollow non-bearing		C 129
Concrete building brick	N,S	C 55

tiples of 8″. It is recommended that the design of concrete block construction be performed to utilize the vertical and horizontal nominal modular dimensions of the units. The width of concrete hollow unit masonry members is determined by the actual size of the block; however, it is designated by its nominal dimension. It is common practice to designate hollow masonry units in terms of the nominal dimensions in the sequence of width, height, and length.

Hollow unit masonry walls are constructed by lapping the units in the successive vertical courses by $\frac{1}{2}$ the unit length to provide a vertical unobstructed core within the wall. This arrangement of the blocks, known as common bond, is the method most often used in the construction of masonry walls. A stack bond arrangement is constructed when the blocks of each successive vertical course are placed directly over each other and all the vertical cells are aligned. Stack bond construction requires that the wall be tied together with horizontal reinforcement or that open end blocks be used to obtain a bond between the blocks in the mortar head joints. The vertical core formed by the aligned cells of the masonry units provides a space for the reinforcing steel. Vertical cores that contain reinforcement should be not less than 2″ × 3″ in the horizontal dimensions and must be filled with cement grout.

Horizontal and vertical mortar joints of hollow masonry unit construction are $\frac{3}{8}$″ thick. The horizontal surfaces of masonry units or bed joints have a full mortar cover on the exterior face shells of the unit and on the horizontal surfaces of cross webs of the blocks that are used to form vertical cores to contain reinforcement. Any mortar that may overflow into the vertical core should be removed to allow the cement grout to bond to the interior surface of the core and to permit proper clearance for the reinforcing bars. The vertical mortar joints, or the head joints, as they are called, should be filled with mortar for a distance in from the exterior face of the block equal to, but not less than, the thickness of the face shell of the block. The mortar joints between concrete blocks should be made straight and with a uniform thickness. Excess mortar that is squeezed from a joint as the result of positioning a block should be removed with a trowel. Concrete block is usually constructed with either a flush joint or a concave joint. Concave joints are made by tooling the surface of the joint so that the mortar will be compressed into the joint rather than its being removed from the surface; however, mortar joints should only be tooled when the mortar is partially set and still plastic. Mortar joints should be made flush with the shell surface of the block when it is to be covered with a coat of plaster.

Horizontal and vertical cells of concrete blocks that contain reinforcing bars, ledger bolts, and other inserts must be filled solid with cement grout. Grout pours should not be made in lifts higher than 4′, and the grout should be well consolidated in place. Cleanout holes should be provided at the bottom of all cores that are to be filled, and 1 hr. should elapse between successive 4′ high pours. Grout pours should be stopped $1\frac{1}{2}$″ below the top of the cell of a course so that a key for the succeeding pour can be formed. The grouting of masonry beams should be performed in one operation, and the tops of cells that are to remain unfilled with grout, which are directly below the cells of masonry beams, should be covered with metal lath to prevent grout leakage. Masonry beams and lintels should be supported in place with shores. It is recommended that these shores remain in place for a period of not less than seven days or at least until the member is capable of supporting its own weight and any construction load that may occur.

Many of the concrete hollow unit masonry details that are presented in this chapter are also presented as details constructed of reinforced grouted solid brick masonry. This type of construction consists of two or more wythes of bricks which are bonded together by mortar joints and by a solid vertical core of cement grout between the wythes. The grout, the mortar, and the bricks are bonded to each other to the degree that they will react as a unit and as a monolithic material. The structural capacity of reinforced grouted brick depends on the design, the quality of the mortar and the cement grout, the durability of the brick, the size and the compressive strength of the brick, and the water absorption factor. Bricks are manufactured by molding mixtures of clay and shale into oblong shapes which are then burned in a kiln to harden them. Although the materials used in the manufacture of bricks may vary, depending on the local source of supply, they must conform to the requirements of A.S.T.M. Spec. No. C 62. Also, bricks are produced with a variety of surface textures and colors, depending on the manufacturer. Table 2 lists the standard nominal modular sizes of bricks.

Mortar joints of reinforced solid masonry should be straight and have a uniform thickness, the bricks should be laid in full head and bed joints, the joints should be not less than ½″ thick, and any excess mortar should be removed from the surface of the bricks after they are in place. Bricks should be dampened at the time they are laid to prevent the initial suction of the surface from removing too much water from the mortar or grout. The suction of the brick is a prime factor in creating the bond between the mortar and the brick; a mortar mixture that contains a larger amount of water will have a higher bond strength. When a brick member or wall is constructed of more than two wythes, the interior brick should be floated into place with not less than ¾″ of cement grout surrounding it. Brick walls are constructed with the same common bond or stack bond arrangement as that described for hollow unit masonry. The method of tooling mortar joints and the time required for soffit shoring of hollow unit masonry previously described also apply to reinforced grouted brick masonry.

Detail 35 (a) and (b) show the minimum clearance required for reinforcing steel in grouted brick masonry walls. The grout space should not be less than the sum of the diameters of the vertical and horizontal reinforcing bars plus ¼″ clear on each side of the bars. It is recommended that the grout space be not less than 2″ thick for reinforced grouted masonry and not less than ¾″ thick in unreinforced grouted masonry. Reinforcing steel in grouted masonry is used in the same way as it is in reinforced concrete, that is, to resist tension or compression in a member that reacts to external forces by bending, compressing, or a combination of both. Therefore, the basic methods and principles of structural design of grouted masonry are the same as those for reinforced concrete, except that the allowable working stresses are adjusted for the masonry materials used. The reinforcing bars are embedded within the grout space between the brick tiers and are spaced and covered with grout or masonry as specified in Table 3. Grout pour requirements vary in different building codes; however, it is generally recommended that low-lift grout pours not exceed 12″ in height for vertical core widths less than 2″ It is also recommended that high-lift grout pours may be made when the vertical core is 2″ or more in width and the grout lift height should not exceed 48 times the core width for mortar grout, or 64 times the core width for pea gravel grout, or a maximum height of 12′0″. High-lift grout pours also require that the exterior tiers of the wall be tied together with rectangular ties of No. 9 gauge wire, 4″ wide by 2″ long and spaced 24″ o.c. horizontally and 16″ o.c. vertically, for walls using common bond bricks. When a stacked bond of the bricks is used, the wire ties between the exterior tiers should be spaced 24″ o.c. horizontally and 12″ o.c. vertically. The grout pours for reinforced masonry should be terminated 1½″ below the top of the bricks to form a key for the succeeding pour. The Uniform Building Code of the International Conference of Building Officials recommends that high-lift grout pours do not exceed 4°0″. Also, grout should be consolidated by vibrating or pudlling to achieve a bond with the reinforcing bars and masonry.

Grout and mortar are a mixture of water,

Table 2. Standard Nominal Modular Sizes of Brick

Thickness, in.	Face dimensions in the wall	
	Height, in.	Length, in.
4	2	12
4	2⅔	8
4	2⅔	12
4	4	8
4	4	12
4	5⅓	8
4	5⅓	12

Table 3. Minimum Reinforcing Steel Spacing and Cover in Grouted Masonry

Location of reinforcment in masonry	*Distance*
Maximum spacing between reinforcing bars in walls	48″
Minimum spacing between parallel bars	1 bar diameter or not less than 1″
Minimum cover of reinforcing bars at the bottom of foundations	3″
Minimum cover of reinforcing bars in vertical members exposed to weather or soil	2″
Minimum cover of reinforcing bars in columns and at the bottoms or sides of girders or beams	1½″
Minimum cover of reinforcing bars in interior walls	¾″

Note: Reinforcing bars that are perpendicular to each other are permitted point contact at their intersection.

portland cement, sand aggregate, and lime putty or hydrated lime. The proportions of the mixture are specified by the volume ratio of the ingredients. Mortar and grout are mixed in the ratio of 1 part portland cement, ¼ to ½ part lime putty or hydrated lime, and 2¼ to 3 parts damp, loose sand. When the grout is to be poured in a space that is 3″ or more in width, the mix may be 1 part portland cement, 2 to 3 parts damp, loose sand, and 2 parts pea gravel or ⅜″ aggregate. Grout should have a fluid consistency when it is pumped into place; however, it should not be so fluid that the constituent aggregates of the mixture will segregate. Mortar should be used within 1½ hr. after it is mixed, but it may be retempered with water during that time to maintain a workable plasticity. Each building code specifies requirements for mortar and grout mixtures; however, the general requirements for mortar should conform to A.S.T.M. Spec. No. C 270 and A.S.T.M. Spec. No. C 476 for grout. Since mortar and grout are composed of water and cement, the temperature conditions at the time it is placed can be critical. The temperature of masonry should be maintained at 50°F for 24 hr. for mixes composed of high early strength cement and 72 hr. for mixes composed of regular types of cement.

Acceptable construction of hollow unit masonry and grouted brick masonry depends a great deal on the quality of the workmanship in the field. Problems will often occur concerning the placement of reinforcing steel, the locations and the details for embedded conduits or pipes, the placing and aligning of construction joints, and the waterproofing of wall surfaces The engineer should try to anticipate these conditions so that the construction will proceed without delays or extra costs. The size and location of reinforcing bars should be accurately dimensioned on the drawings; vertical reinforcement should be secured in place at the top and bottom and at intervals not exceeding 192 bar diameters apart; horizontal reinforcing bars should be placed in bond beam or channel blocks in hollow unit masonry construction and within grout widths or as shown in Details 30 to 32 (b); and reinforcing bars should be lapped 30 bar diameters or not less than 24″. Possible problems caused by locating pipes or conduits that pass through a masonry wall or are embedded within a wall can be avoided by coordinating the structural drawings with the requirements of the mechanical and electrical drawings. The location and the method of providing for the pipe or conduit in the wall should be shown on the structural drawings; all pipes that pass through masonary walls should be sleeved with a standard wrought iron pipe to prevent them from bonding with the masonry. Pipes and conduits in cores that are not filled with cement grout are not considered as embedded in the wall. However, when a pipe or conduit is embedded in masonry construction, care should be taken to ensure that it is not located at points of high shear or flexure stress and that it does not reduce the structural value of the wall.

Masonry walls will expand or contract, depending on the climate conditions and the composition of the material. The coefficient of thermal expansion of hollow masonry and brick masonry will vary in relation to the materials used in their manufacture; however, the engineer should consider the necessity of vertical construction joints in walls to allow for thermal growth, particularly in climates where the extreme temperature values will vary over a wide range. There are many recommended types and spacings of vertical wall joints. In general, a vertical wall joint should be sufficient width to permit the free horizontal movement of the wall and also be capable of maintaining the structural capacity of the wall. This can be accomplished by aligning a full height vertical joint through the wall cross section. The joint space is filled with a compressible bituminous or waterproof material, and the horizontal reinforcing bars are lapped across the joint but are not bonded to the grout on one side. The absence of expansion joints in masonry walls can be a source of severe cracking and therefore reduce the structural value of the wall.

Masonry is a comparatively porous material; if it is not waterproofed and it is exposed to normal weather conditions, it will allow a certain amount of moisture to permeate it, especially if the walls are approximately 8″ thick. The moisture that penetrates a wall from the exterior to the interior of the building will leave a white chalklike stain on the interior surface of the masonry. This phenomenon is caused by efflorescence, which occurs when the moisture that passes through the wall evaporates from the interior surface and leaves a white, dry, water-soluble salt residue that is derived from the masonry material. Efflorescence can be prevented by waterproofing the exterior surface of the wall and by sealing any cracks in the mortar joints by retooling them. Other sources of water intrusion in masonry walls can be traced to insert openings or cracks, to exposed grout surfaces at door and window openings, and to the tops of parapet walls. There are many commercial products available for the purpose of waterproofing masonry walls. The recommended methods and frequency of application of these materials are usually specified in their guarantee of performance. All masonry walls that are constructed below grade should be waterproofed with two layers of either asphalt or coal tar pitch material. Materials specifically manufactured for this purpose are also available and should be applied to the exterior surface of the wall in accordance with the manufacturer's recommendations. A definite method should also be provided to remove large amounts of water that may collect adjacent to walls below grade. This can be done either by installing perforated ceramic drain tiles adjacent to the bottom of the wall or by backfilling the area adjacent to the bottom of the wall with a continuous pocket of crushed rock or coarse aggregate. The tiles or crushed rock backfill should be sloped to permit the water to flow away from the wall. Exposed grout surfaces of masonry walls at windows, doors, and parapets should be covered with a glavanized sheet metal flashing. Jamb, sill, and header frames at doors and windows should be pressure caulked to the masonry surface. The layers of the roofing material should extend above the roof onto the surface of the parapet wall and should be either flashed into the wall or extended over the top of the wall and be covered by the parapet sheet metal flashing. Improper moisture protection of masonry walls can be a source of much damage within a building. Small leaks around windows and doors or through parapet copings may cause damage to large areas of finished plaster walls and ceilings and expensive floor covering materials; therefore it is quite important that positive steps be taken in the construction of a building to eliminate the possibility of water leakage.

Reinforced grouted masonry is designed by the same basic methods and principles used to design reinforced concrete. It is also assumed that grouted masonry will react to externally applied loads in the same manner that reinforced concrete will react, that is, that the masonry is not capable of resisting tensile stress, that the tension in a masonry member is resisted only by the reinforcing steel, and that the cement grout and the reinforcing steel are bonded together and will react as an analogous monolithic material. The primary difference between reinforced concrete and reinforced grouted masonry is the working stress values assigned to the grouted masonry by the various building codes or design criteria.

These values are determined by compressive tests on the masonry and are specified as f'_m. The values of f'_m depend on the type of masonry used and on the quality of field workmanship during construction. The values of the allowable unit stresses of masonry acting in compression, shear, and bond are listed in the building codes for masonry construction performed with continuous inspection and without continuous inspection. Since the quality of field workmanship is an important factor for the strength of reinforced grouted masonry, the allowable unit working stresses with continuous inspection are double the allowable stresses that do not require continuous inspection. Valid continuous inspection requires that a registered deputy building inspector be on the job during the execution of all masonry construction and that he or she inspect and verify that the work is being performed in accordance with the building code requirements and that it complies with the engineer's structural design. The building code is not a design manual; it is a specification of the minimum safe design and construction requirements. Engineers, contractors, and field inspectors should bear this fact in mind when they use masonry as a construction material. This statement is not to be construed as meaning that masonry is not a good construction material; rather, it stresses that the quality of the workmanship in the use of masonry as a structural material is a determining factor in the structural strength. No construction is better than the least technically qualified person on the job, regardless of whether that person is designing, drafting, supervising or mixing mortar.

Masonry offers many advantages as a construction material since it is incombustible and has a high thermal insulation value. Depending on the structural requirements, the cost of masonry wall construction can be favorably compared with that of reinforced concrete construction since masonry walls can be constructed without the labor and material costs of concrete forms.

DETAILS

Notes ▪ Drawings ▪ Ideas

8" WIDE WALL

8" HIGH UNITS

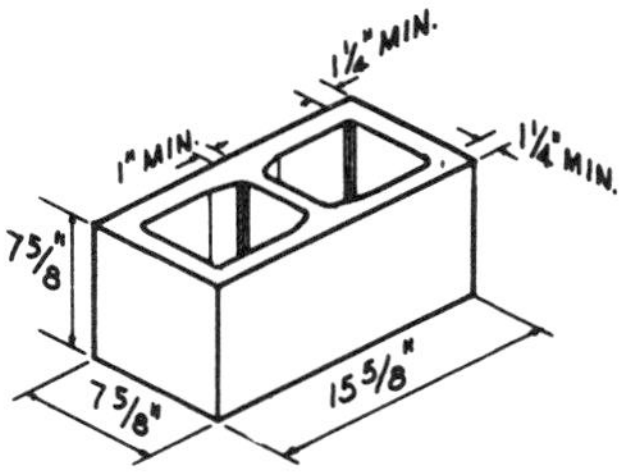

8 x 8 x 16
STANDARD

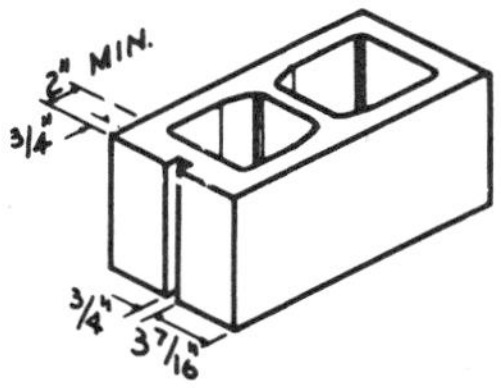

8 x 8 x 16
SASH

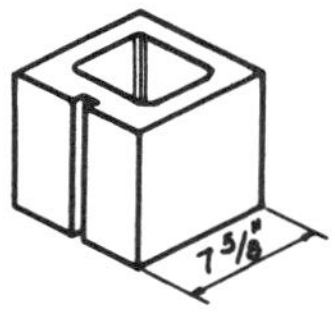

8 x 8 x 8
HALF SASH

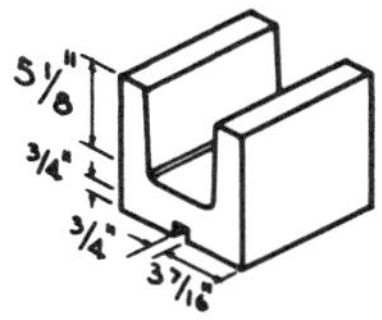

8 x 8 x 8
SASH LINTEL

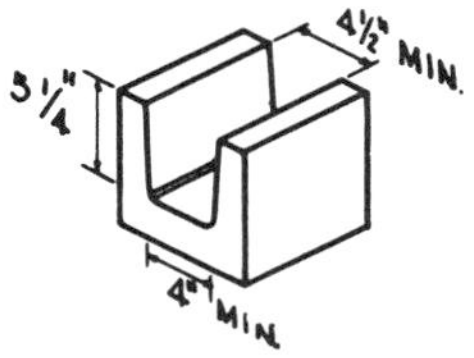

8 x 8 x 8
STANDARD LINTEL

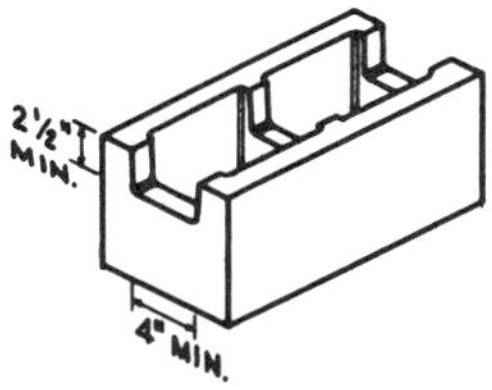

8 x 8 x 16
BOND BEAM

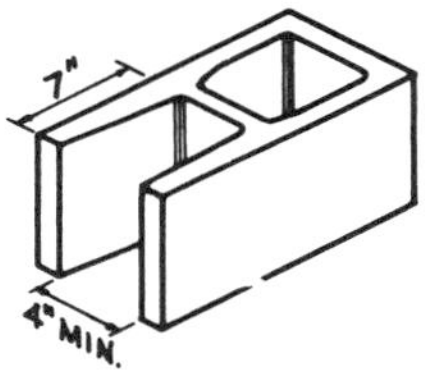

8 x 8 x 16
OPEN END

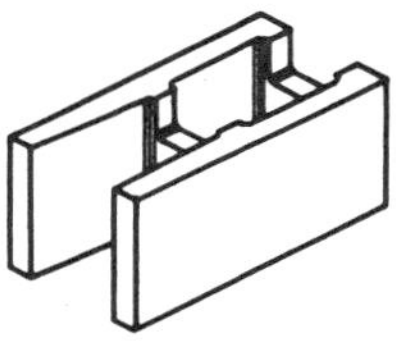

8 x 8 x 16
OPEN END
BOND BEAM

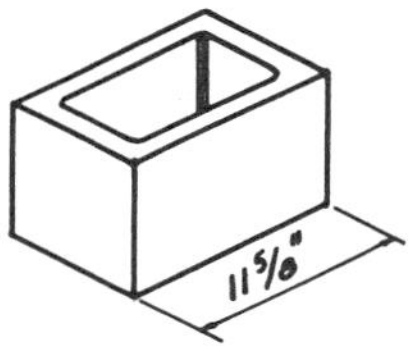

8 x 8 x 12
THREE QUARTER

4" HIGH UNITS

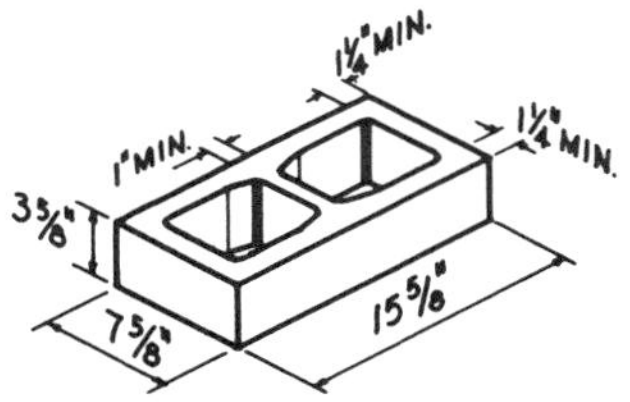

8 x 4 x 16
STANDARD

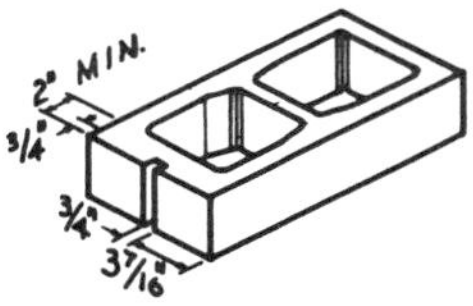

8 x 4 x 16
SASH

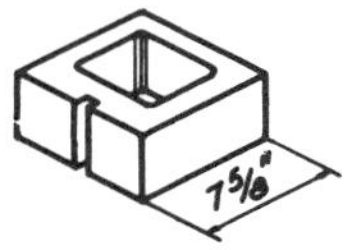

8 x 4 x 8
HALF SASH

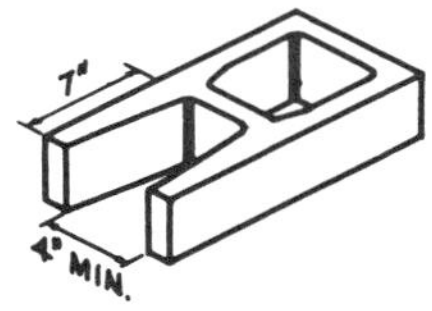

8 x 4 x 16
OPEN END

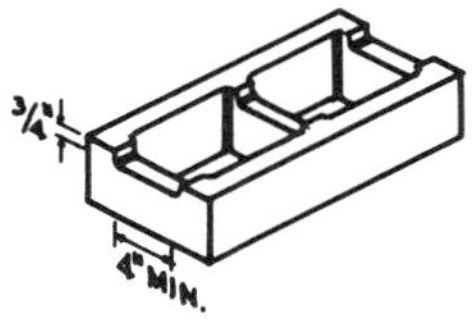

8 x 4 x 16
CHANNEL

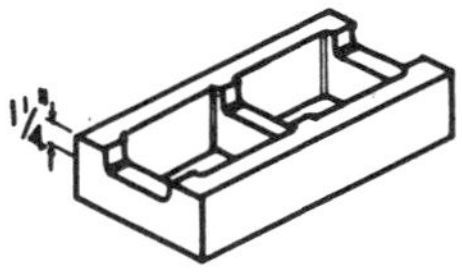

8 x 4 x 16
BOND BEAM

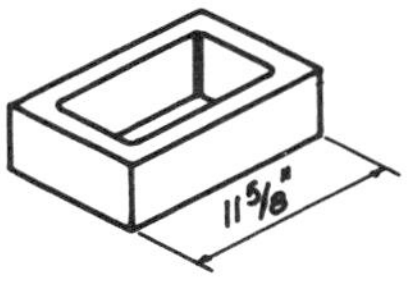

8 x 4 x 12
THREE QUARTER

Notes ▪ Drawings ▪ Ideas

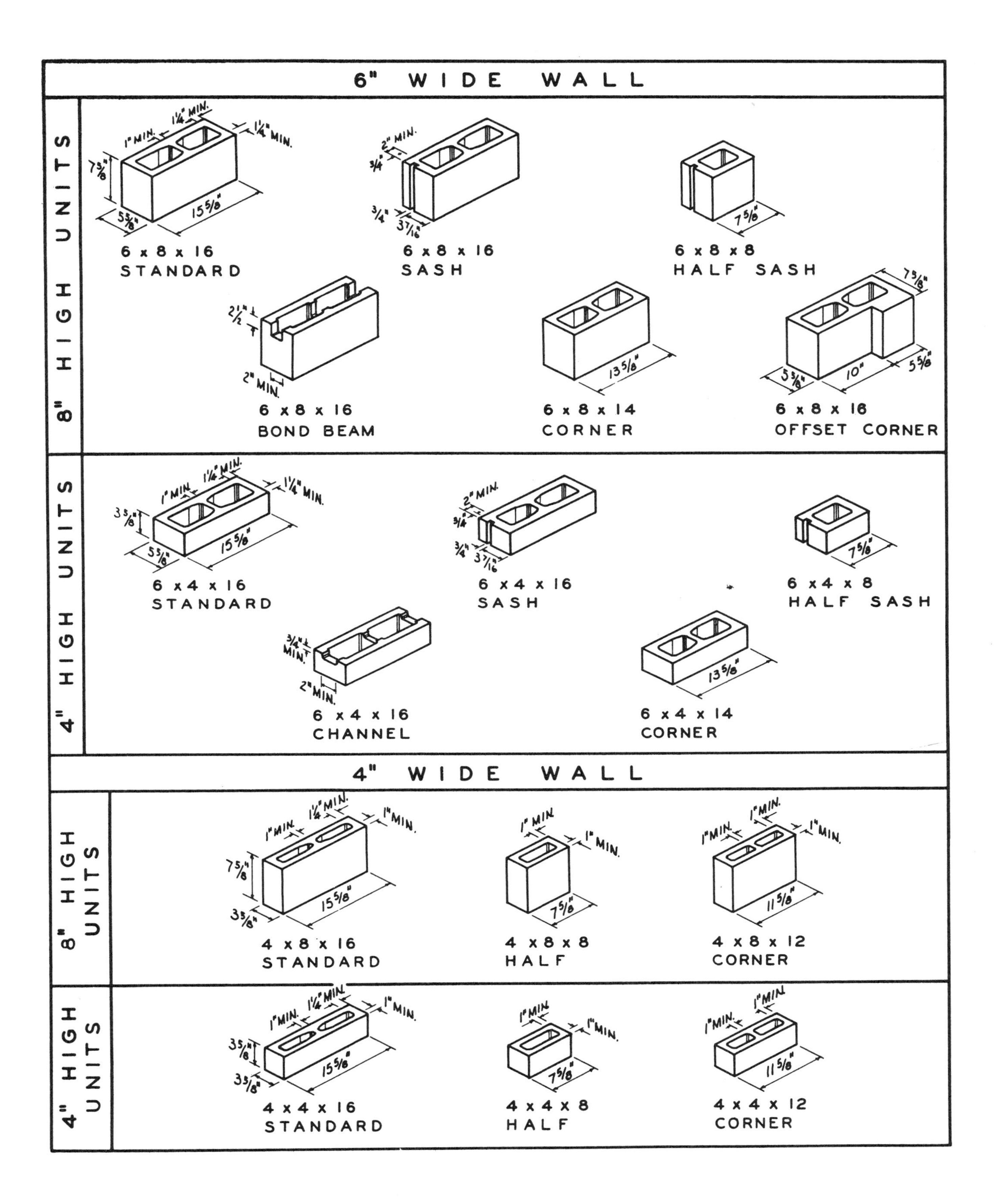

6" WIDE WALL
8" HIGH UNITS
6 x 8 x 16
STANDARD
6 x 8 x 16
SASH
6 x 8 x 8
HALF SASH
6 x 8 x 16
BOND BEAM
6 x 8 x 14
CORNER
6 x 8 x 16
OFFSET CORNER
4" HIGH UNITS
6 x 4 x 16
STANDARD
6 x 4 x 16
SASH
6 x 4 x 8
HALF SASH
6 x 4 x 16
CHANNEL
6 x 4 x 14
CORNER
4" WIDE WALL
8" HIGH UNITS
4 x 8 x 16
STANDARD
4 x 8 x 8
HALF
4 x 8 x 12
CORNER
4" HIGH UNITS
4 x 4 x 16
STANDARD
4 x 4 x 8
HALF
4 x 4 x 12
CORNER

Notes ▪ Drawings ▪ Ideas

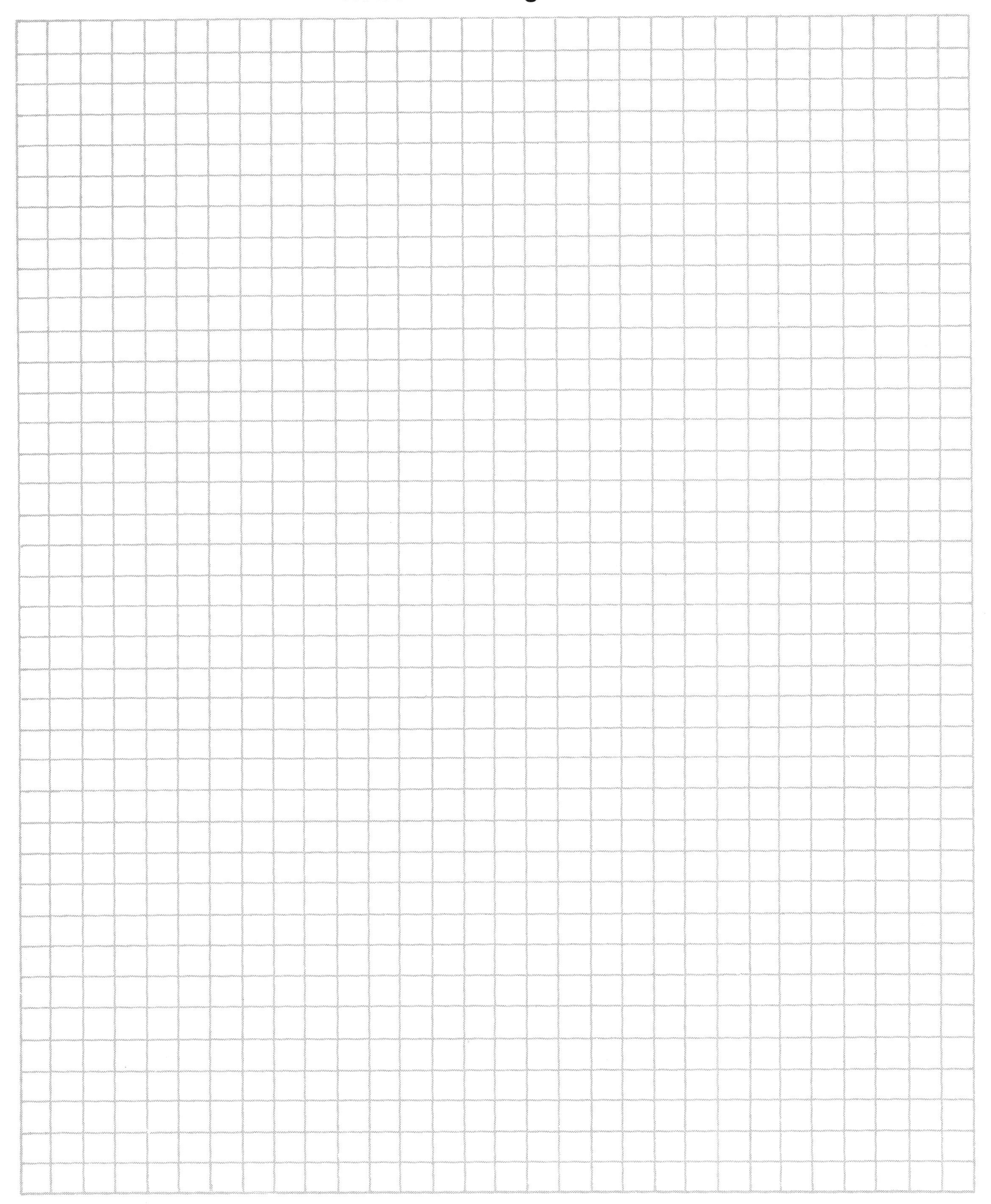

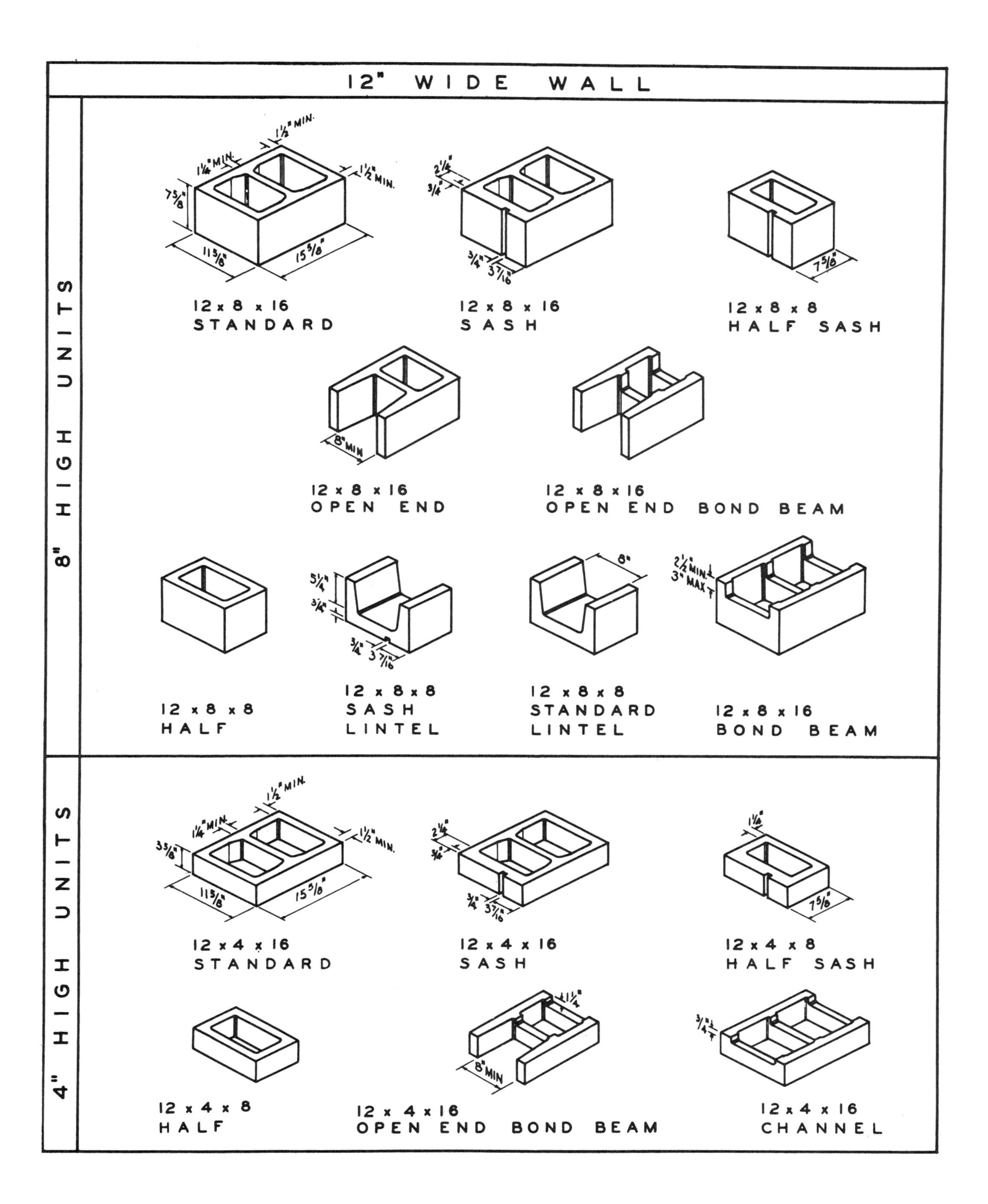
12" WIDE WALL
8" HIGH UNITS
1½" MIN.
1¼" MIN.
1½" MIN.
7⅝"
11⅝"
15⅝"
12 x 8 x 16
STANDARD
2¼"
¾"
¾"
3 7/16"
12 x 8 x 16
SASH
7⅝"
12 x 8 x 8
HALF SASH
8" MIN
12 x 8 x 16
OPEN END
12 x 8 x 16
OPEN END BOND BEAM
5¼"
¾"
¾"
3 7/16"
8"
2½" MIN.
3" MAX.
12 x 8 x 8
HALF
12 x 8 x 8
SASH
LINTEL
12 x 8 x 8
STANDARD
LINTEL
12 x 8 x 16
BOND BEAM
4" HIGH UNITS
1½" MIN.
1¼" MIN.
1½" MIN.
3⅝"
11⅝"
15⅝"
12 x 4 x 16
STANDARD
2¼"
¾"
¾"
3 7/16"
12 x 4 x 16
SASH
1¼"
7⅝"
12 x 4 x 8
HALF SASH
1¼"
8" MIN
¾"
12 x 4 x 8
HALF
12 x 4 x 16
OPEN END BOND BEAM
12 x 4 x 16
CHANNEL

Notes ▪ Drawings ▪ Ideas

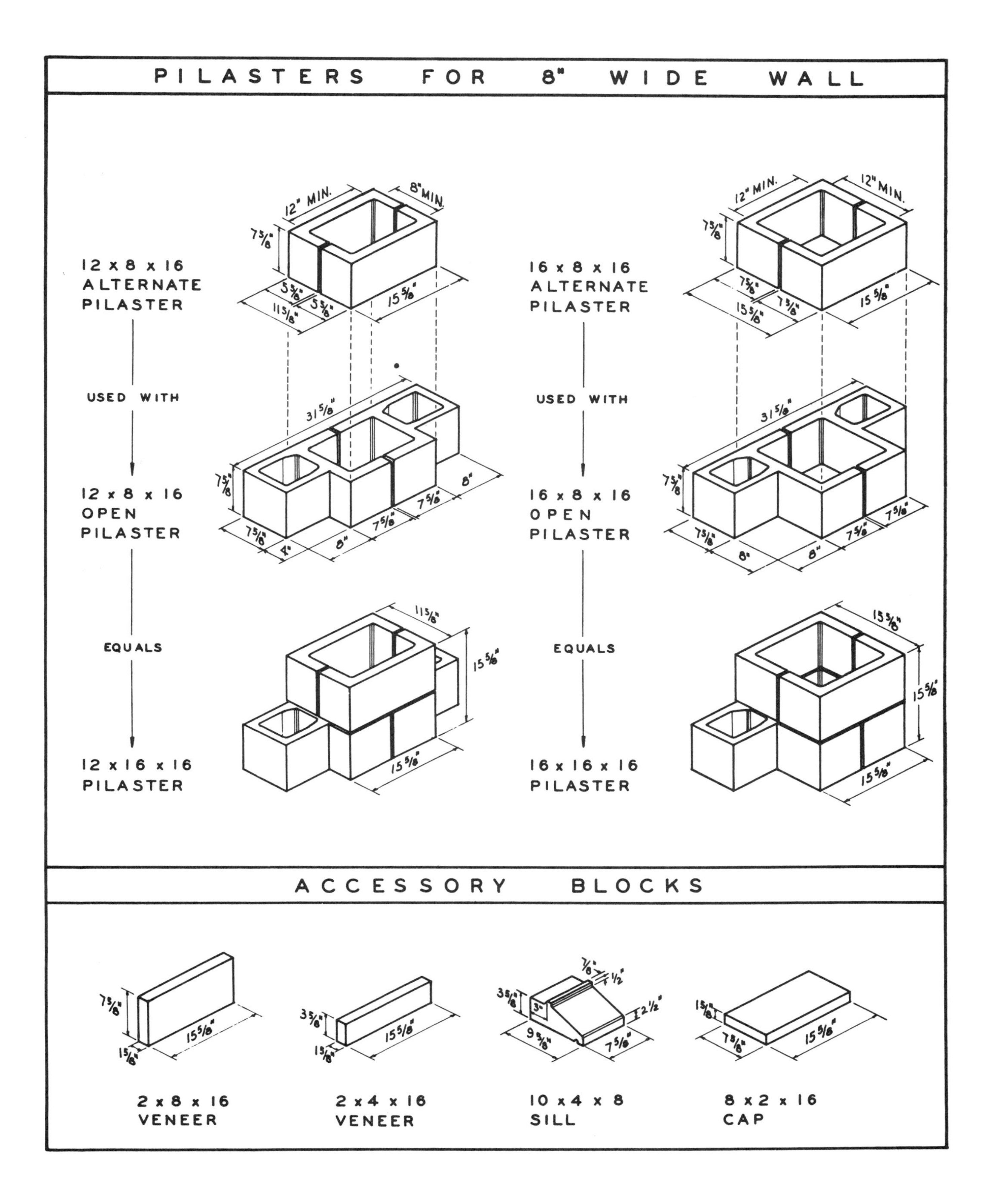

PILASTERS FOR 8" WIDE WALL
12 x 8 x 16
ALTERNATE
PILASTER
USED WITH
12 x 8 x 16
OPEN
PILASTER
EQUALS
12 x 16 x 16
PILASTER
16 x 8 x 16
ALTERNATE
PILASTER
USED WITH
16 x 8 x 16
OPEN
PILASTER
EQUALS
16 x 16 x 16
PILASTER
12" MIN.
8" MIN.
12" MIN.
12" MIN.
7 5/8"
15 5/8"
11 5/8"
31 5/8"
ACCESSORY BLOCKS
2 x 8 x 16
VENEER
2 x 4 x 16
VENEER
10 x 4 x 8
SILL
8 x 2 x 16
CAP

Notes ▪ Drawings ▪ Ideas

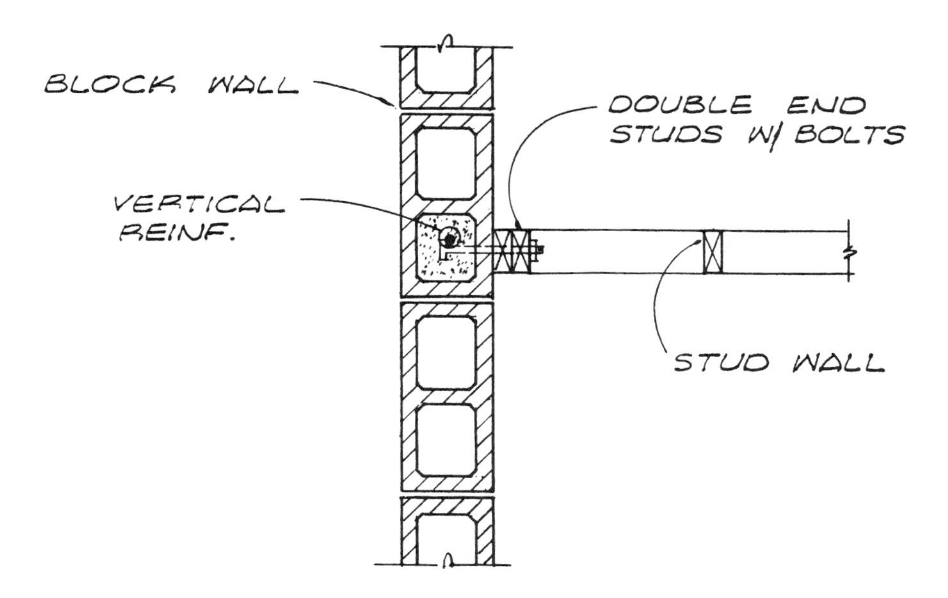

Detail 1. A plan section of a concrete block masonry wall intersected by a wood stud wall. The wood stud wall is nailed to a wood nailer that is bolted to the masonry wall as shown. One vertical reinforcing bar is placed in the wall at the line of the wall intersection.

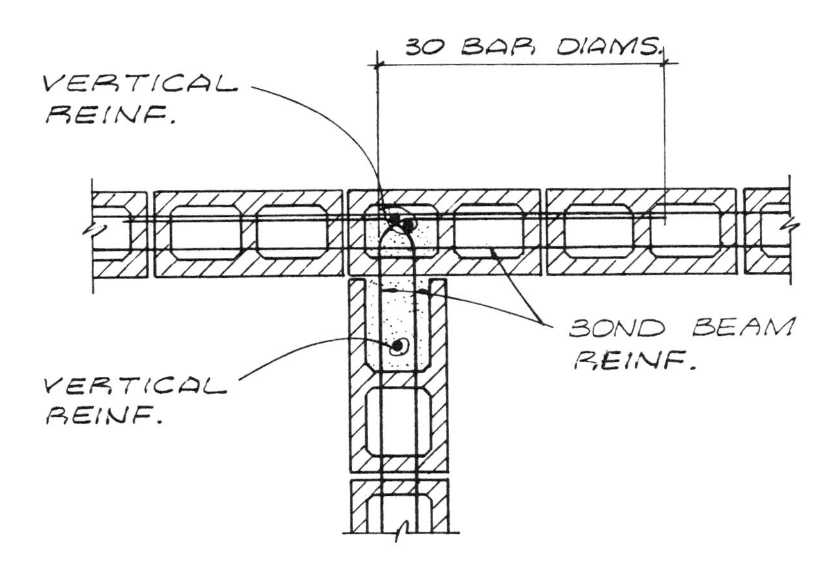

Detail 2. A plan section of the intersection of concrete block masonry wall bond beams. The vertical reinforcing bars are added to the wall at the line of the wall intersection. The bond beam reinforcing steel is lapped 30 bar diameters or a minimum of 24″. Fill all masonry cells that contain reinforcement.

Notes ▪ Drawings ▪ Ideas

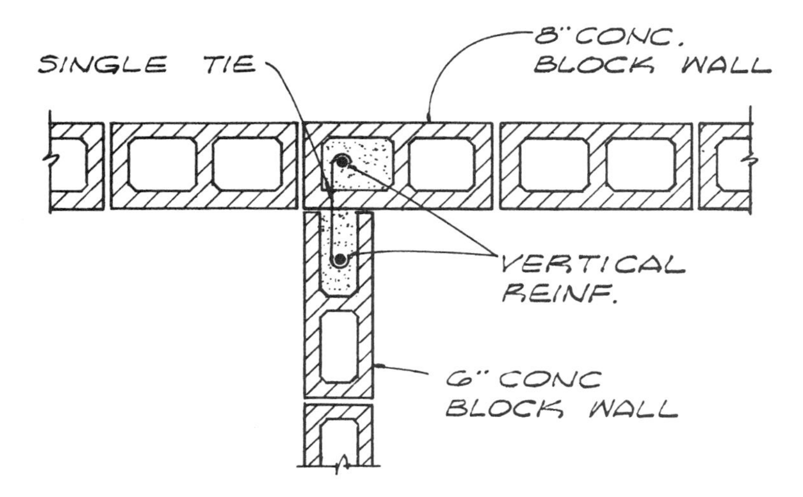

Detail 3. A plan section of the intersection of concrete block masonry walls. The walls are connected by dowels hooked to vertical reinforcing bars in each wall. Fill all masonry cells that contain reinforcement.

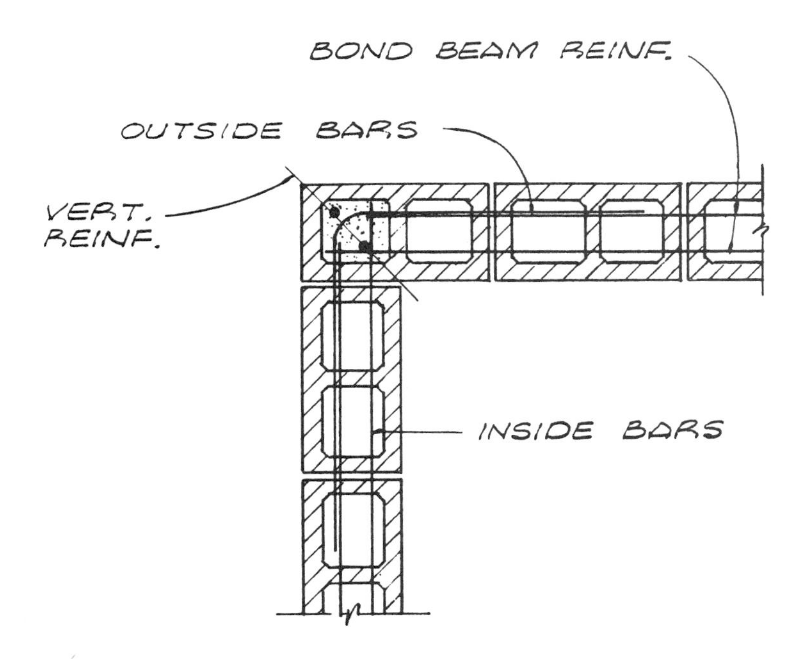

Detail 4. A plan section of the corner of a concrete block masonry wall. Two vertical reinforcing bars are added in the corner cell of the wall. The bond beam reinforcing bars at the outside face of the wall are lapped as shown. The reinforcing steel lap is 30 bar diameters or a minimum of 24″. Fill all masonry cells that contain reinforcement.

Notes ▪ Drawings ▪ Ideas

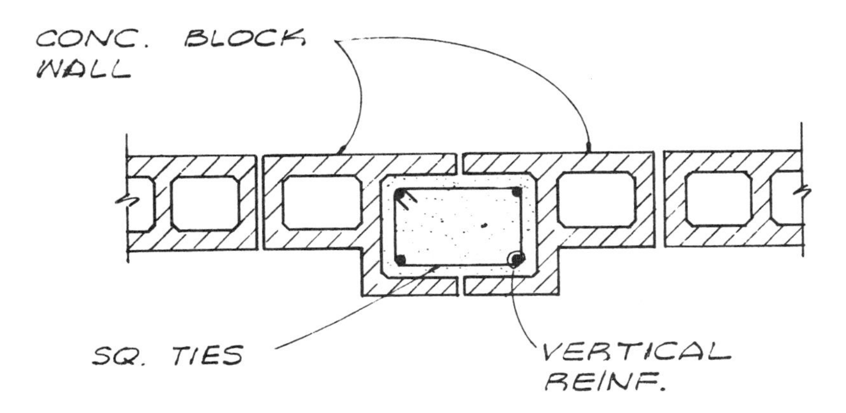

Detail 5(a). A plan section of a concrete block masonry wall rectangular pilaster. The pilaster is reinforced with a minimum of four vertical bars tied together in the same manner as is required for a rectangular concrete column. Fill all masonry cells that contain reinforcement.

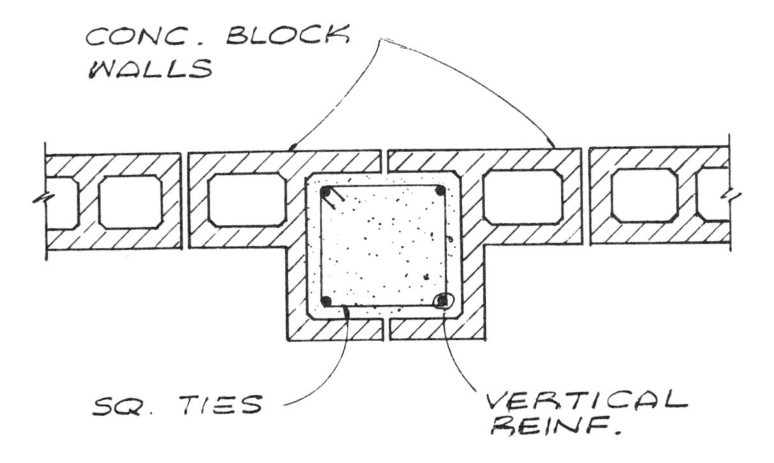

Detail 5(b). A plan section of a concrete block masonry wall square pilaster. The pilaster is reinforced with a minimum of four vertical bars tied together in the same way as is required for a rectangular concrete column. Fill all masonry cells that contain reinforcement.

Notes ▪ Drawings ▪ Ideas

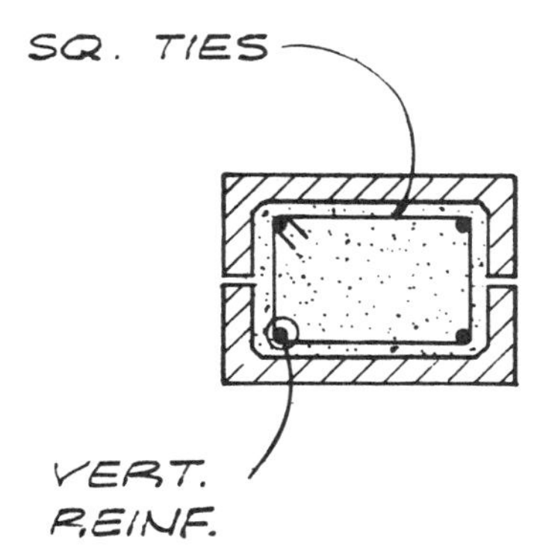

Detail 6(a). A plan section of a concrete block masonry rectangular column. The column is reinforced with a minimum of four vertical bars tied together in the same way as is required for a rectangular concrete column. Fill all masonry cells that contain reinforcement.

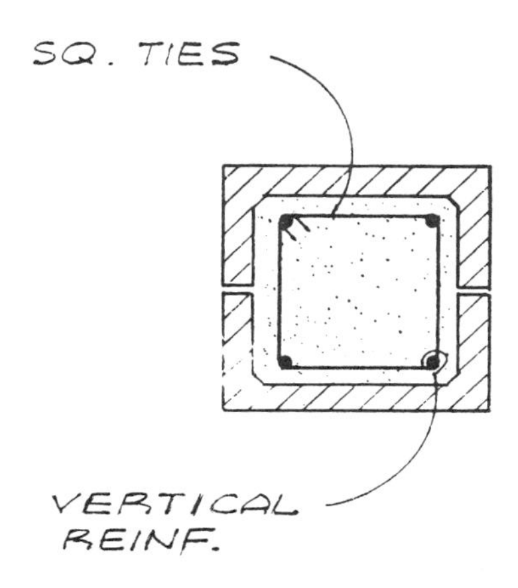

Detail 6(b). A plan section of a concrete block masonry square column. The column is reinforced with a minimum of four vertical bars tied together in the same way as is required for a rectangular concrete column. Fill all masonry cells that contain reinforcement.

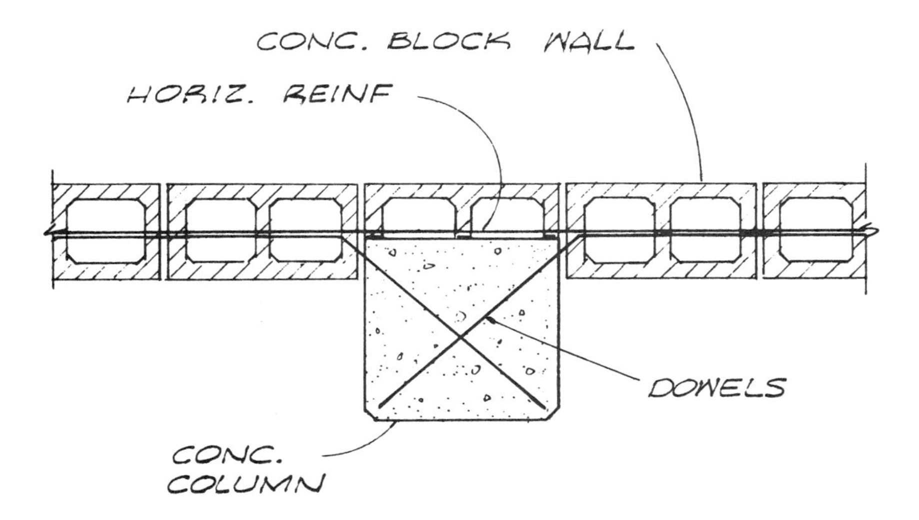

Detail 7. A plan section of a poured concrete column and a concrete block masonry wall. The wall is connected to the column by extending the horizontal reinforcing bars of the wall into the column as shown. The vertical reinforcing bars and ties of the poured column are not shown.

Notes ▪ Drawings ▪ Ideas

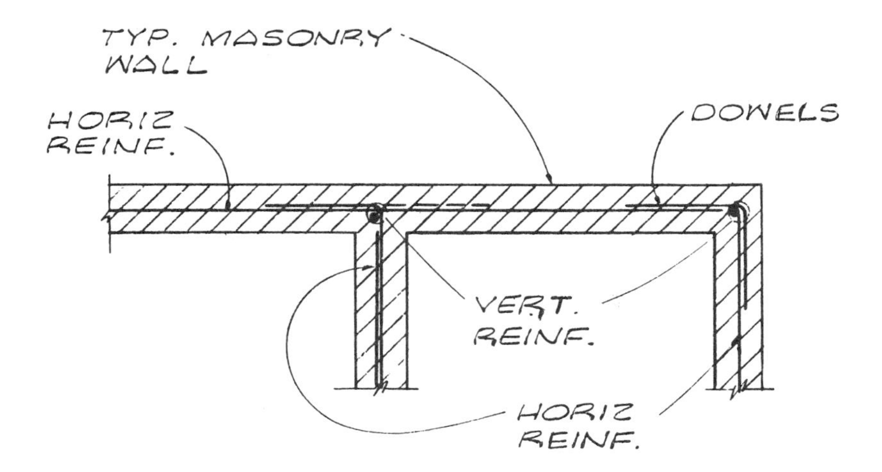

Detail 8. A plan section of a corner and of an intersection of masonry walls. The walls are connected by lapping the horizontal reinforcing bars as shown. The bars are lapped 30 bar diameters or a minimum of 24".

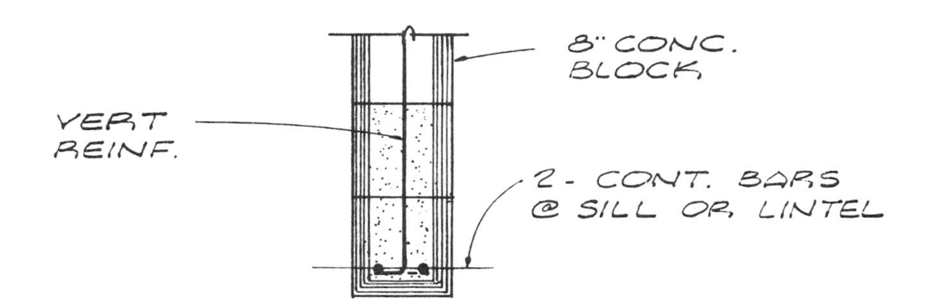

Detail 9(a). A section of a concrete block masonry wall lintel. The size and number of reinforcing bars in the lintel depend on the span and the load and are determined by calculation. The horizontal reinforcing bars extend 24" past the edge of the opening. The depth of the lintel is determined by the number of masonry cells filled with grout. All cells that contain reinforcing bars are filled with grout.

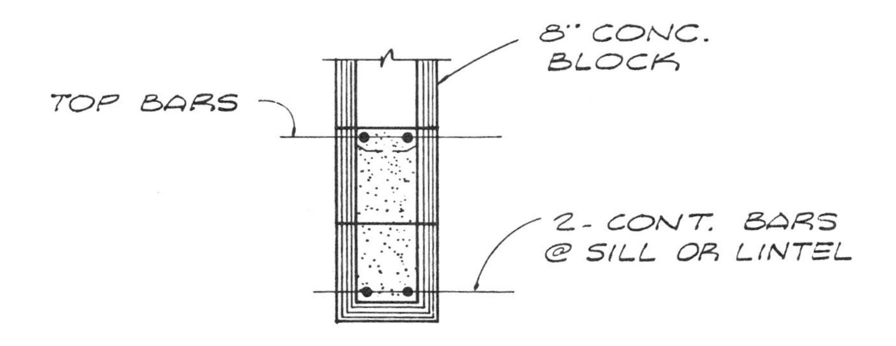

Detail 9(b). A section of a concrete block masonry wall lintel with reinforcing bars at the top and bottom. The size and number of reinforcing bars in the lintel depend on the span and the load and are determined by calculation. The horizontal reinforcing bars extend 24" past the edge of the opening. All cells that contain reinforcing bars are filled with grout. See Detail 10(a).

Notes ▪ Drawings ▪ Ideas

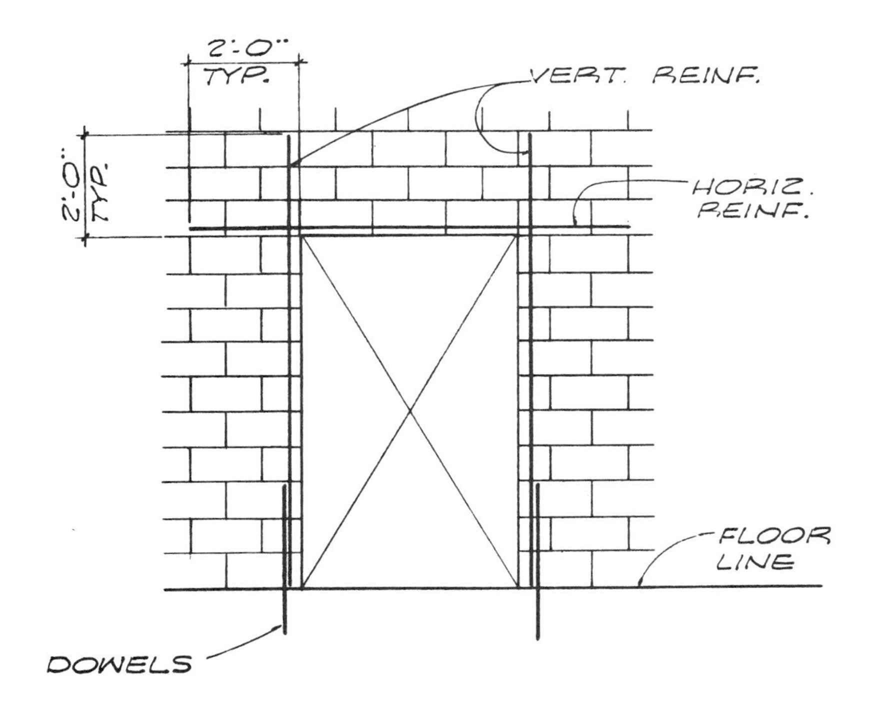

Detail 10(a). An elevation of an opening in a concrete block masonry wall. Vertical and horizontal reinforcing bars are added at the edge of the opening and extend 24″ past the edge of the opening. See Details 9 (a) and (b) for lintel sections; see Detail 10 (d) for a masonry jamb section.

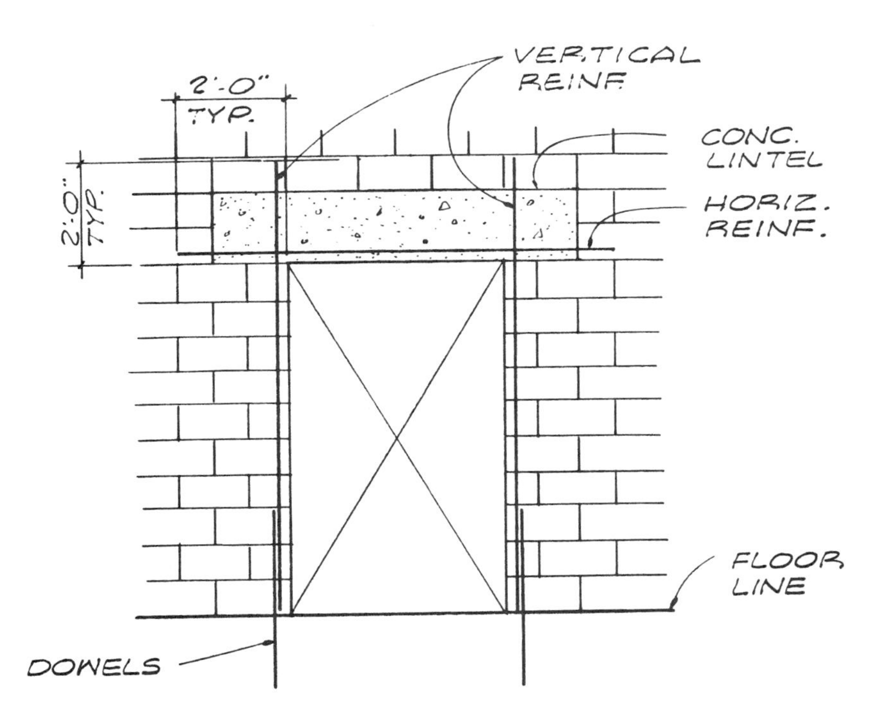

Detail 10(b). An elevation of an opening in a concrete block masonry wall. Vertical and horizontal reinforcing bars are added at the edges of the opening and extend 24″ past the edge of the opening. The lintel is a poured concrete beam. See Detail 10(d) for a section of the lintel and for a masonry jamb section.

Notes ▪ Drawings ▪ Ideas

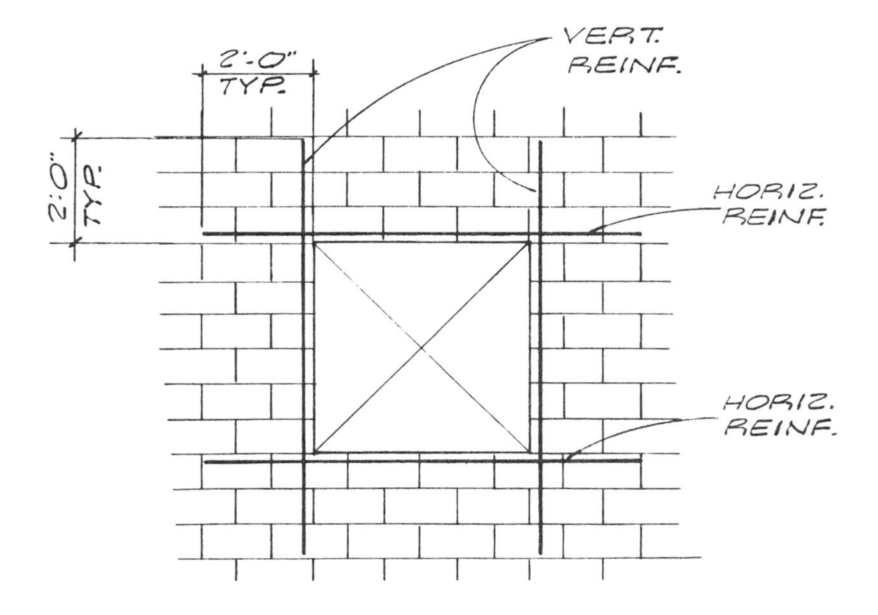

Detail 10(c). An elevation of an opening in a concrete block masonry wall. Vertical and horizontal reinforcing bars are added at the edge of the opening and extend 24″ past the edge of the opening. See Details 9(a) and (b) for masonry lintel sections; see Detail 10(d) for a masonry jamb section.

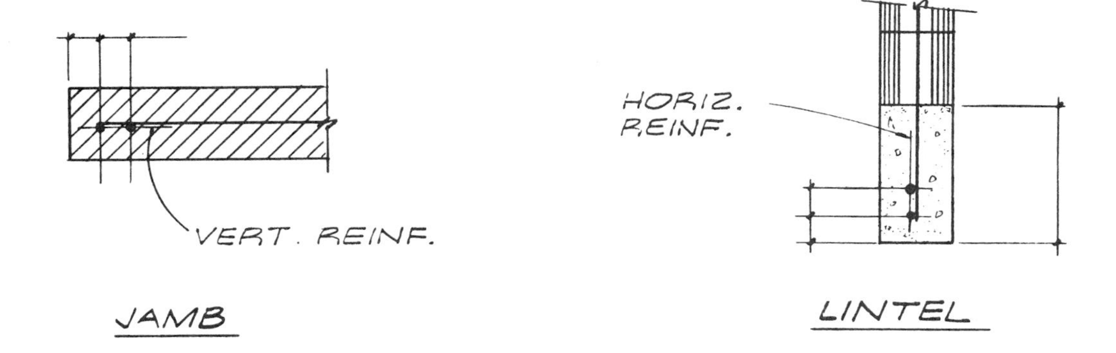

Detail 10(d). A section of a concrete lintel in a concrete block masonry wall, and a section of a jamb for an opening in a concrete block masonry wall.

Notes ▪ Drawings ▪ Ideas

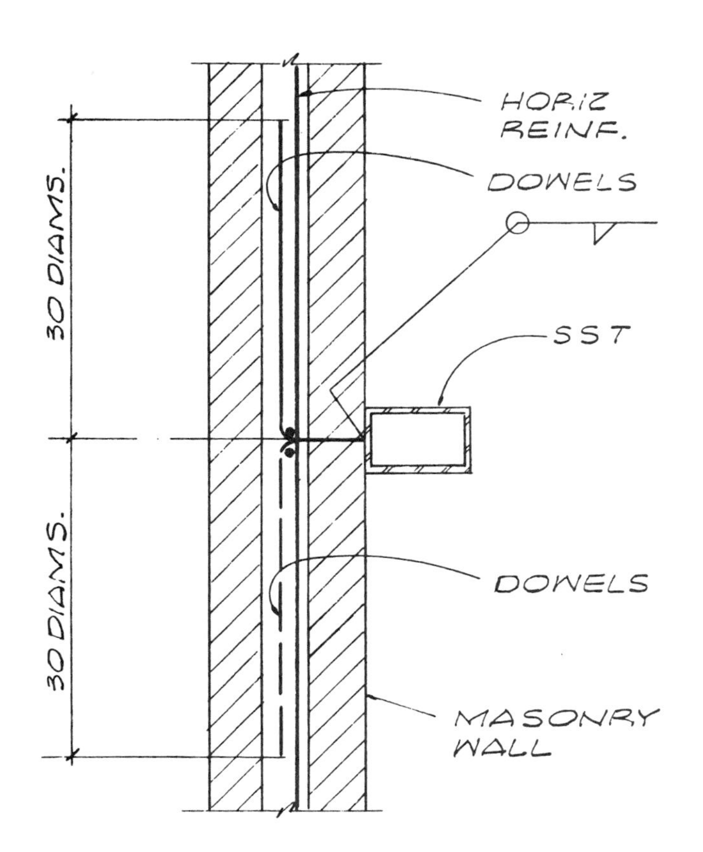

Detail 11. A plan section of a masonry wall and a structural steel tube column. The steel column is connected to the masonry wall by welding horizontal reinforcing dowels at the back face of the column and bending them into the masonry wall. A vertical reinforcing bar is added at the bend of the horizontal dowels. The size and spacing of the dowels into the masonry wall depend on the column forces to be resisted. The reinforcing dowels lap 30 bar diameters into the wall or not less than 24″.

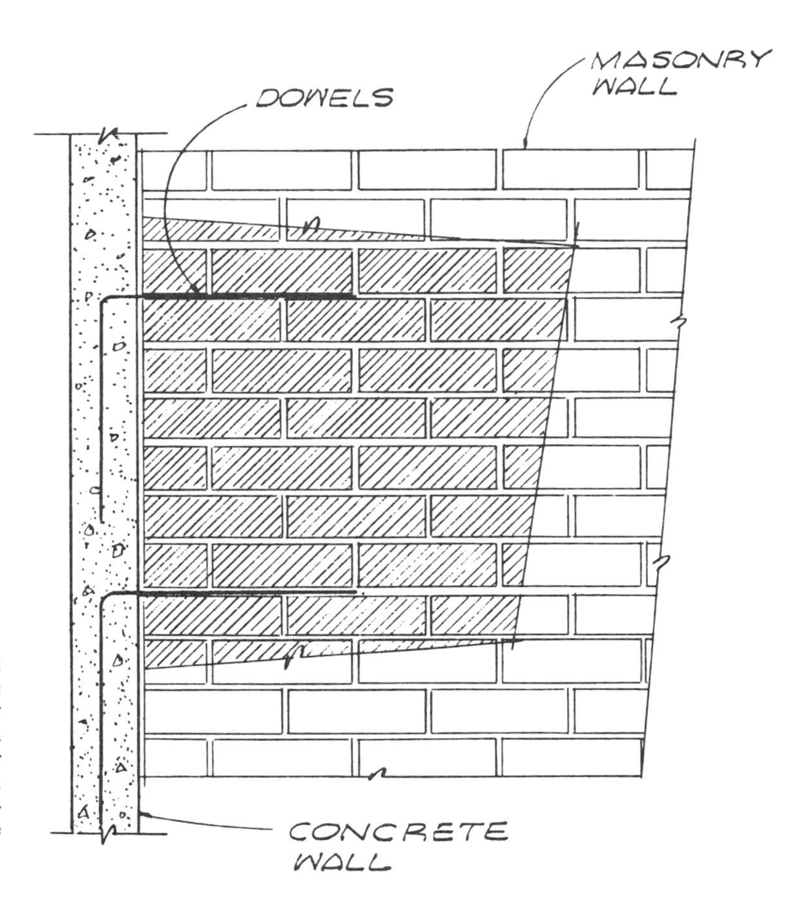

Detail 12. A section of a concrete wall and an elevation of an intersecting masonry wall. The masonry wall is connected to the concrete wall by reinforcing dowels as shown. The dowel lap length is 30 bar diameters into the masonry and 40 bar diameters into the concrete or not less than 24″. The size and spacing of the reinforcing dowels depend on the lateral forces to be resisted by the walls.

Notes ▪ Drawings ▪ Ideas

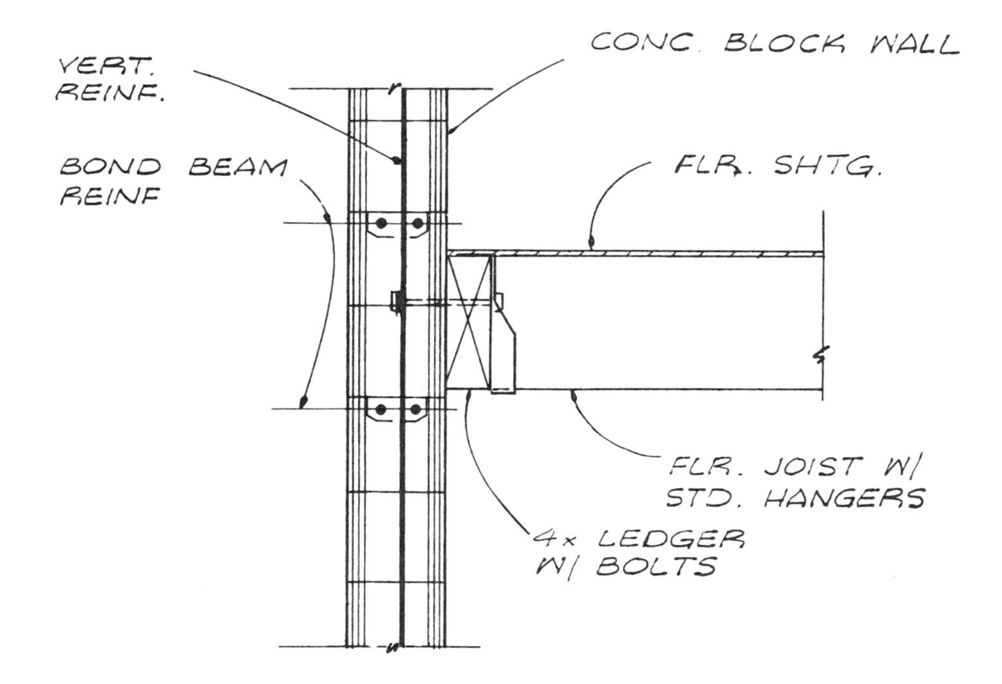

Detail 13. A section of a concrete block masonry wall supporting wood floor joists. The joists are connected to a 4" wide wood ledger by standard joist hangers. The ledger is connected to the masonry wall by bolts. The size and spacing of the ledger bolts are determined by calculation. The horizontal reinforcing bars act as a bond beam in the masonry wall.

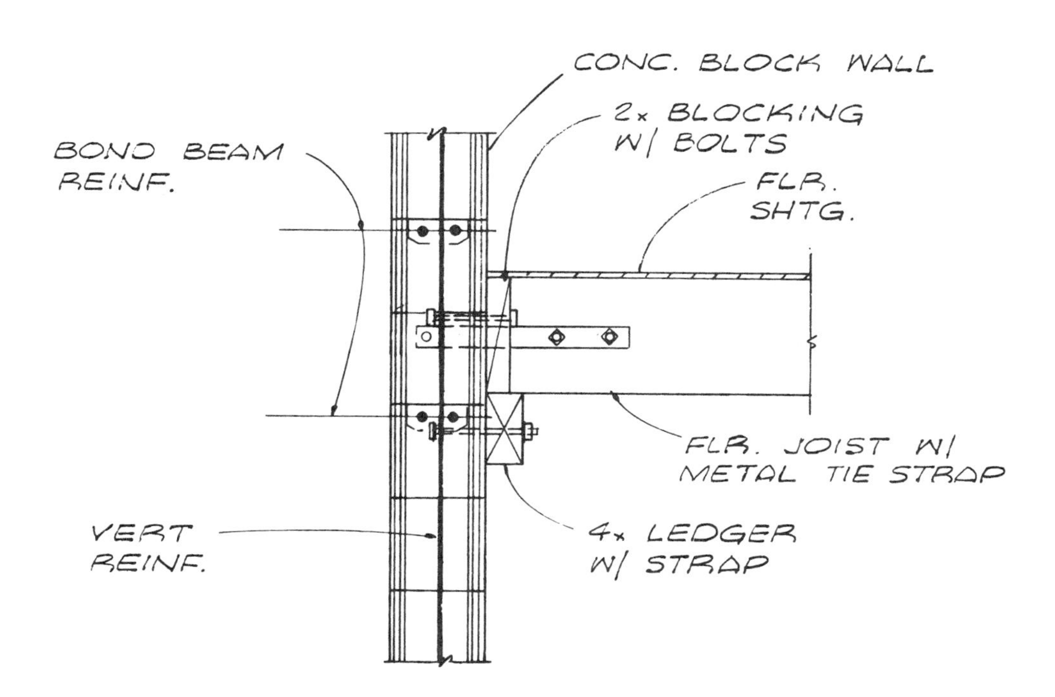

Detail 14. A section of a concrete block masonry wall supporting wood floor joists. The joists bear on a 4" wide wood ledger which is bolted to the masonry wall. The size and spacing of the ledger bolts are determined by calculation. The joist continuous blocking is bolted to the masonry wall. A metal tie strap spaced at 4'0" o.c. connects the floor to the masonry. The horizontal reinforcing bars act as a bond beam in the masonry wall.

Notes ▪ Drawings ▪ Ideas

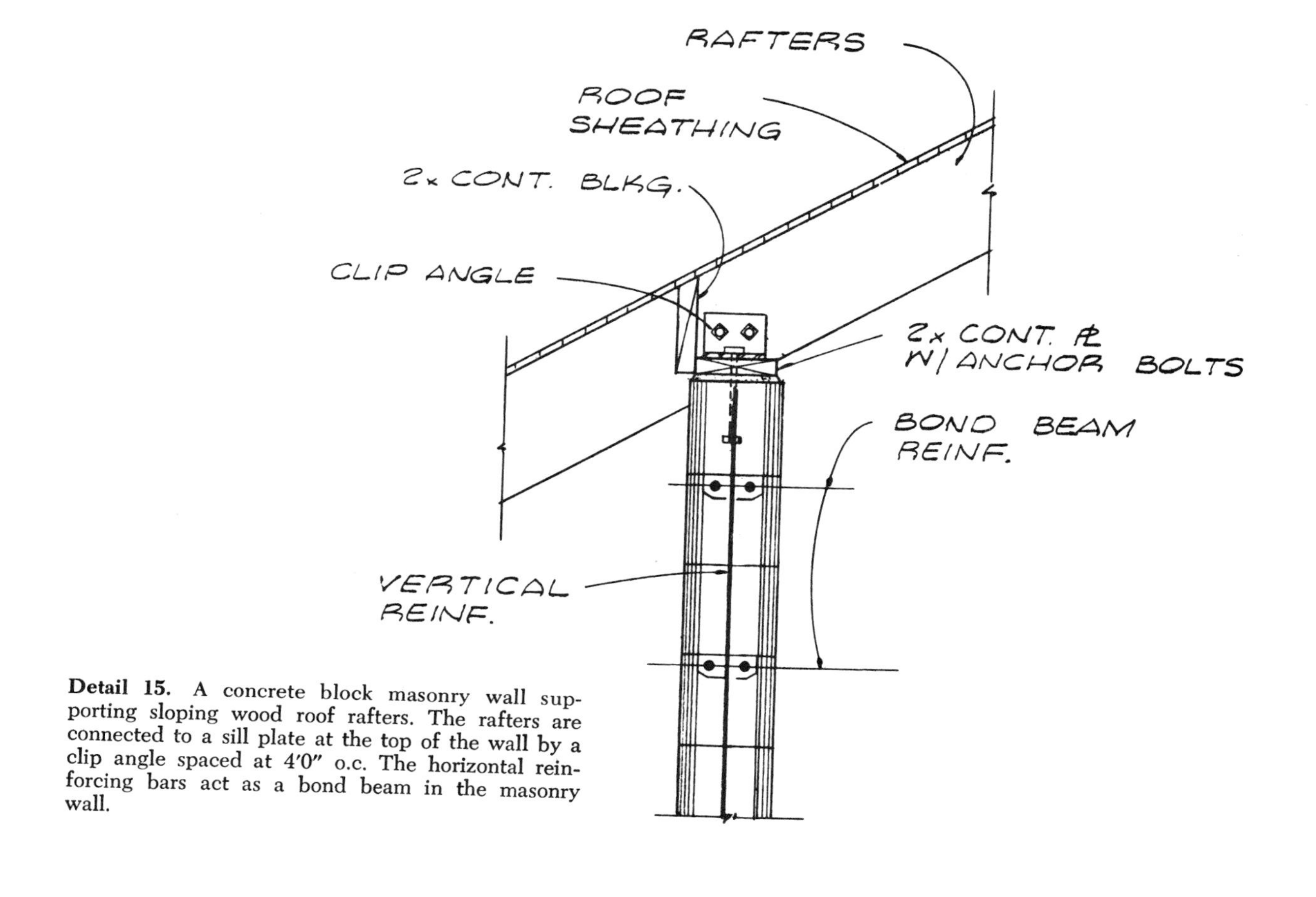

Detail 15. A concrete block masonry wall supporting sloping wood roof rafters. The rafters are connected to a sill plate at the top of the wall by a clip angle spaced at 4′0″ o.c. The horizontal reinforcing bars act as a bond beam in the masonry wall.

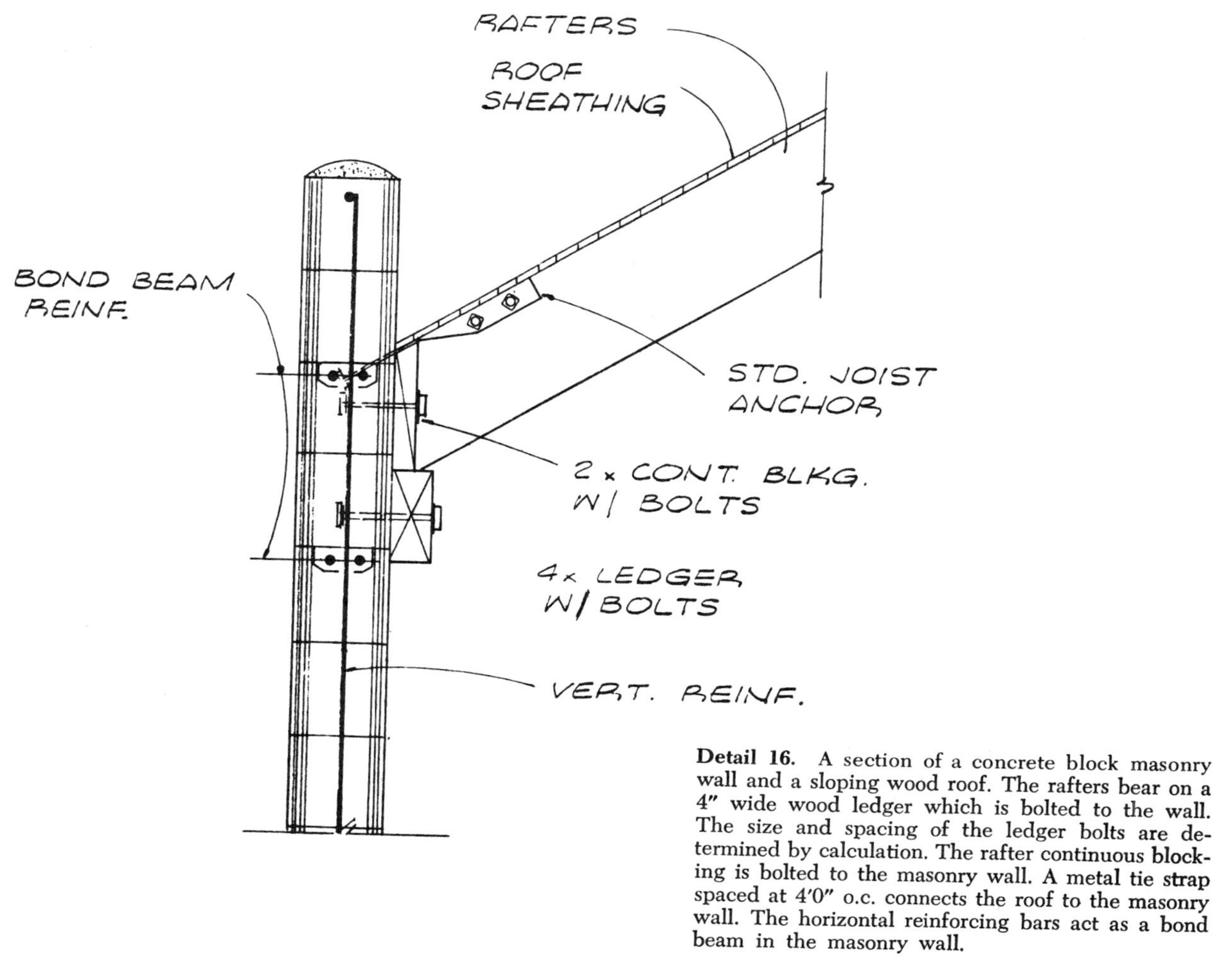

Detail 16. A section of a concrete block masonry wall and a sloping wood roof. The rafters bear on a 4″ wide wood ledger which is bolted to the wall. The size and spacing of the ledger bolts are determined by calculation. The rafter continuous blocking is bolted to the masonry wall. A metal tie strap spaced at 4′0″ o.c. connects the roof to the masonry wall. The horizontal reinforcing bars act as a bond beam in the masonry wall.

Notes ▪ Drawings ▪ Ideas

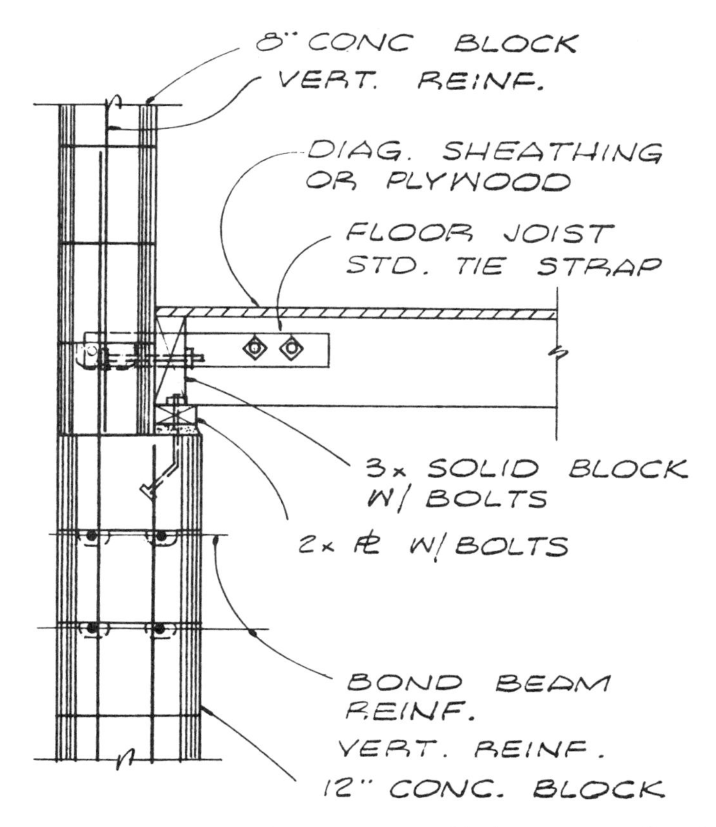

Detail 17(a). A section of a 12" concrete block masonry wall and an 8" concrete block masonry wall supporting wood floor joists. The joists bear on a 4" wide wood plate which is bolted to the top of the 12" wall as shown. The joist continuous blocking is bolted to the 8" wall. A metal tie strap spaced at 4'0" o.c. connects the floor to the wall.

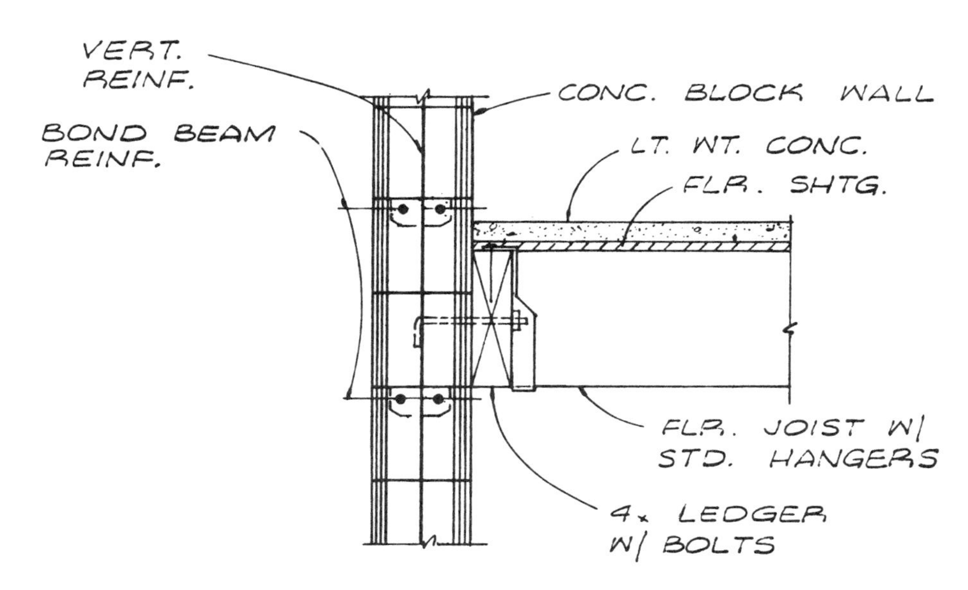

Detail 17(b). A section of a concrete block masonry wall, wood floor joists, and light-weight concrete over a wood floor. The floor joists are connected to a 4" wide wood ledger by standard joist hangers. The ledger is bolted to the masonry wall. The size and spacing of the ledger bolts are determined by calculation. The floor sheathing is nailed to the ledger to transfer the floor diaphragm stress into the wall. The horizontal reinforcing bars act as a bond beam in the masonry wall.

Notes ▪ Drawings ▪ Ideas

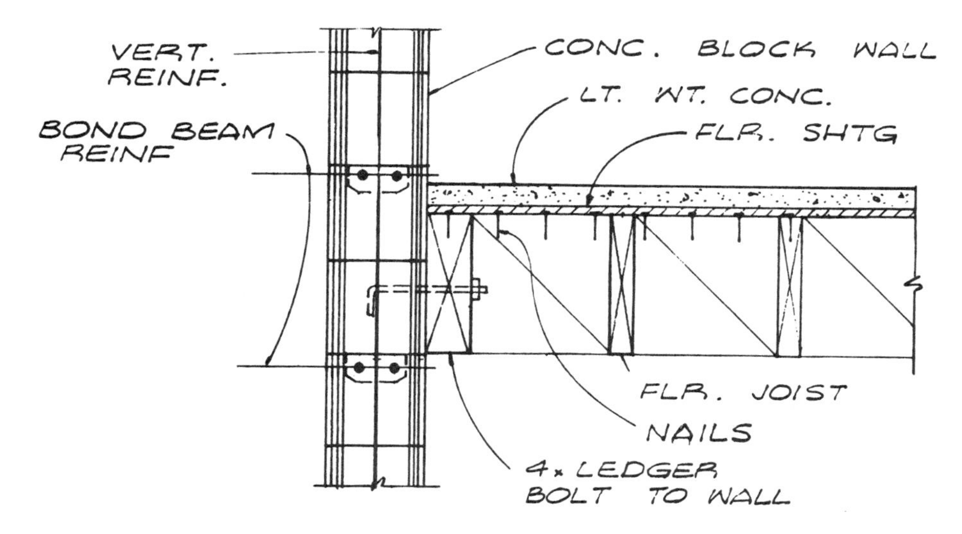

Detail 18(a). A section of a concrete block masonry wall, floor joists, and light-weight concrete over a wood floor. The floor joists span parallel to the wall. The floor sheathing is nailed to a 4" wide wood ledger to transfer the floor diaphragm stress into the masonry wall. The diaphragm stress may require that the perimeter floor joists be blocked and nailed as shown. The ledger is bolted to the masonry wall. The size and spacing of the ledger bolts are determined by calculation. The horizontal reinforcing bars act as a bond beam in the masonry wall.

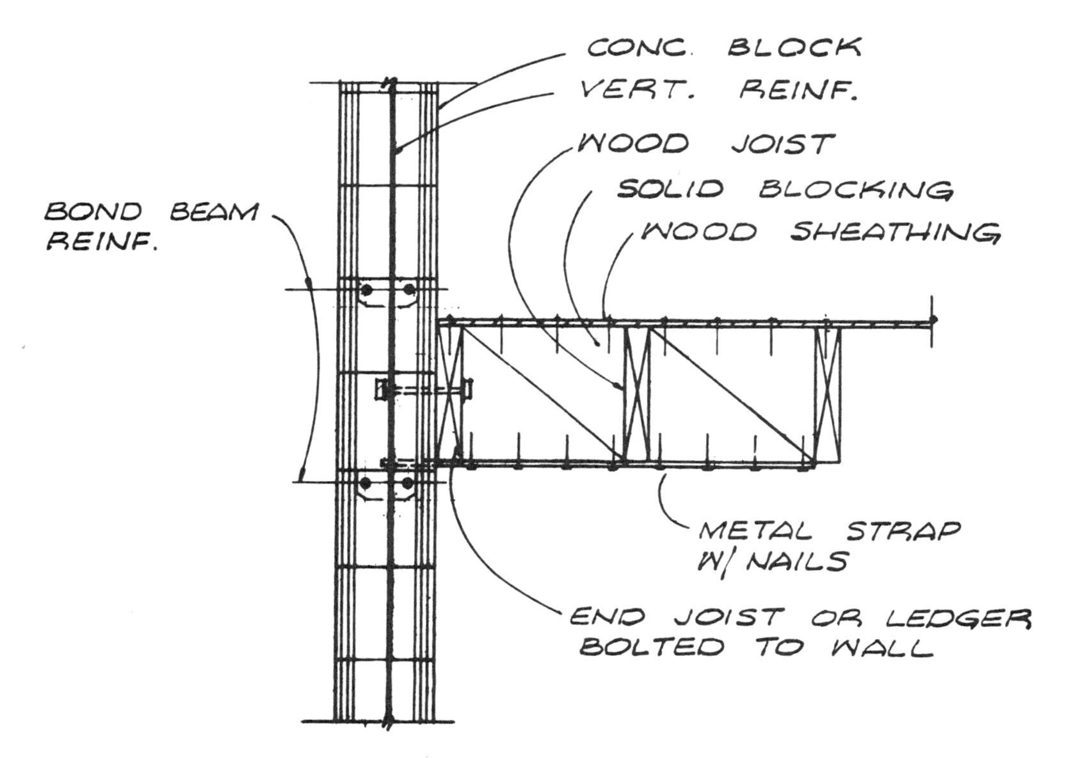

Detail 18(b). A section of a concrete block masonry wall, floor joists, and floor sheathing. The floor joists span parallel to the wall. The floor sheathing is nailed to a 4" wide wood ledger to transfer the floor diaphragm stress into the masonry wall. The diaphragm stress may require that the perimeter floor joists be blocked and nailed as shown. The ledger is bolted to the masonry wall. The size and spacing of the ledger bolts are determined by calculation. A metal tie strap spaced at 4'0" o.c. connects the floor to the masonry wall. The horizontal reinforcing bars act as a bond beam in the wall.

Notes ▪ Drawings ▪ Ideas

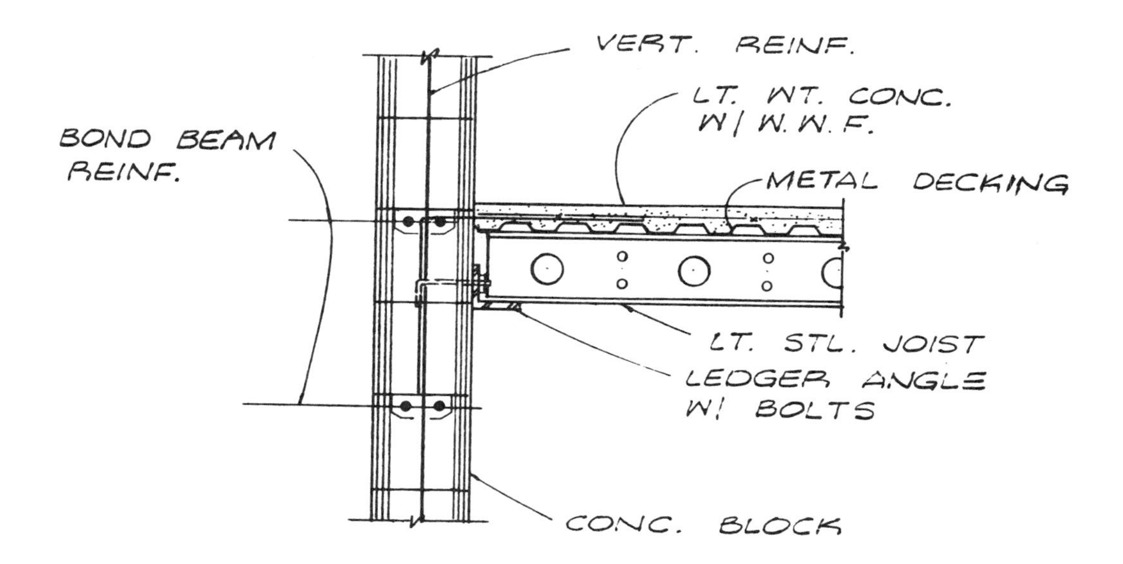

Detail 19. A section of a concrete block masonry wall, light steel joists, a continuous ledger angle, and metal decking with light-weight concrete. The steel joists bear on and are welded to the ledger angle which is bolted to the masonry wall. The size and spacing of the angle bolts are determined by calculation. The light-weight concrete slab is connected to the wall with reinforcing dowels. The horizontal reinforcing bars act as a bond beam in the masonry wall.

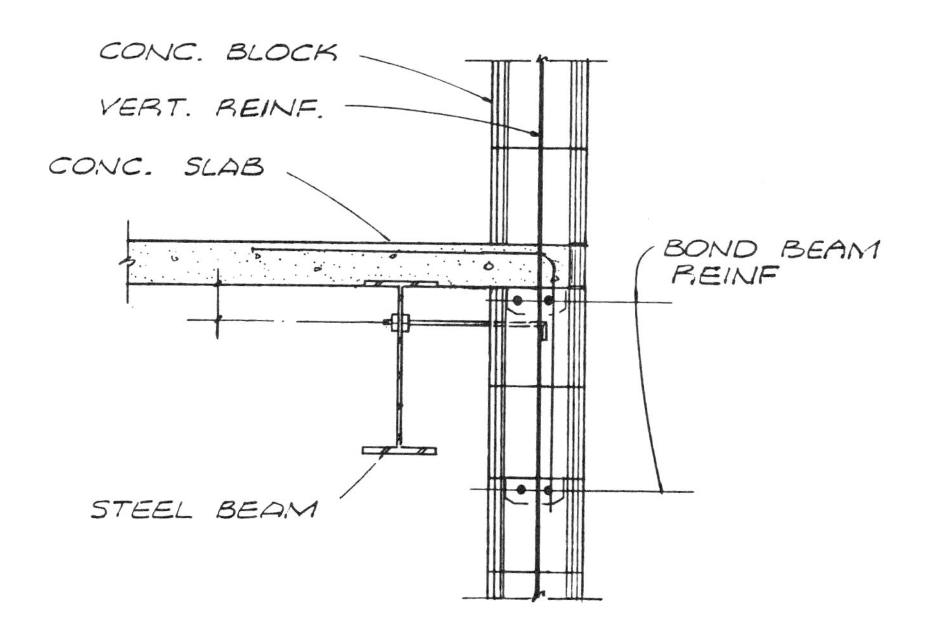

Detail 20. A section of a concrete block masonry wall and a concrete slab. The slab is supported by a steel beam. The slab diaphragm stress is transferred to the masonry wall by reinforcing dowels. Equally spaced rods between the web of the steel beam and the masonry wall restrain the beam laterally. The horizontal reinforcing bars act as a bond beam in the masonry wall.

Notes ▪ Drawings ▪ Ideas

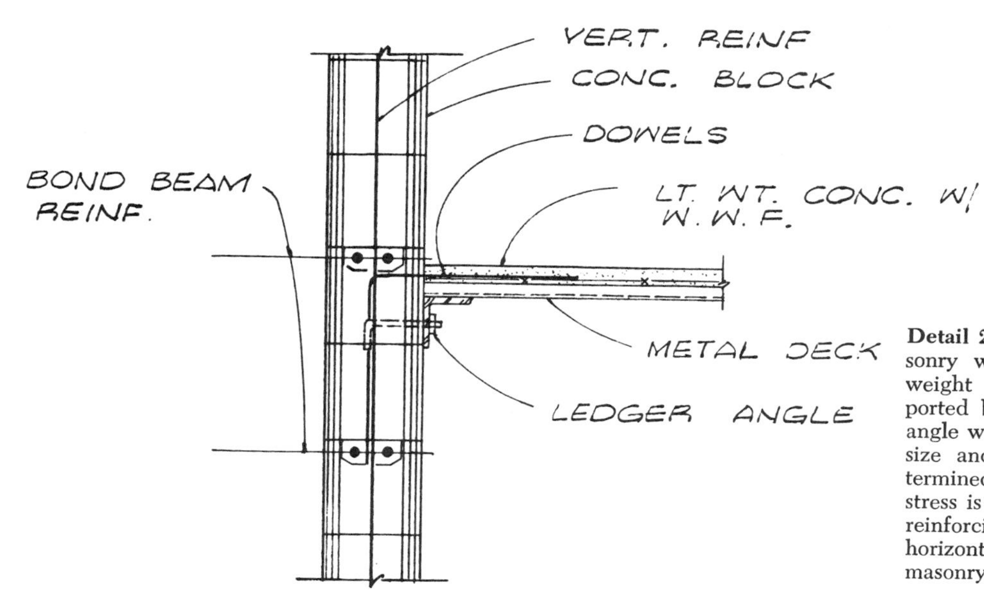

Detail 21. A section of a concrete block masonry wall and a metal decking with lightweight concrete. The metal decking is supported by and welded to a continuous ledger angle which is bolted to the masonry wall. The size and spacing of the angle bolts are determined by calculation. The floor diaphragm stress is transferred to the masonry wall by the reinforcing dowels and the ledger angle. The horizontal bars act as a bond beam in the masonry wall.

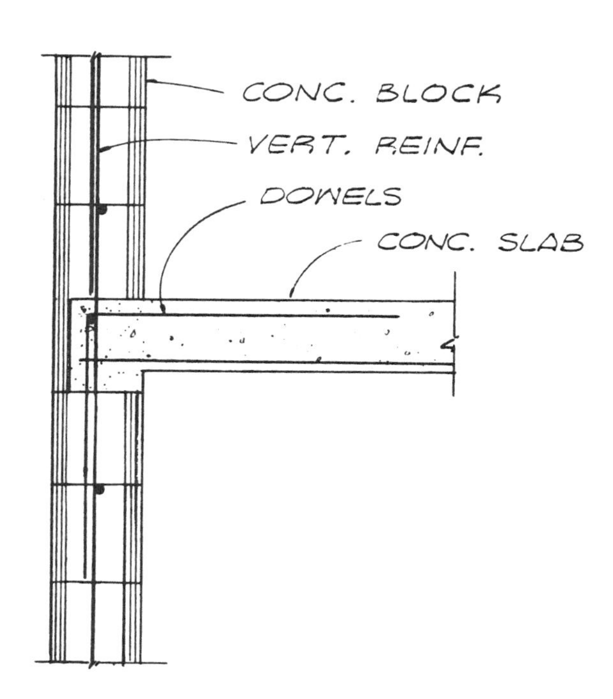

Detail 22. A section of a concrete block masonry wall and a poured concrete slab. The slab is cast into the wall for vertical support. The slab diaphragm stress is transferred to the masonry wall by the reinforcing dowels.

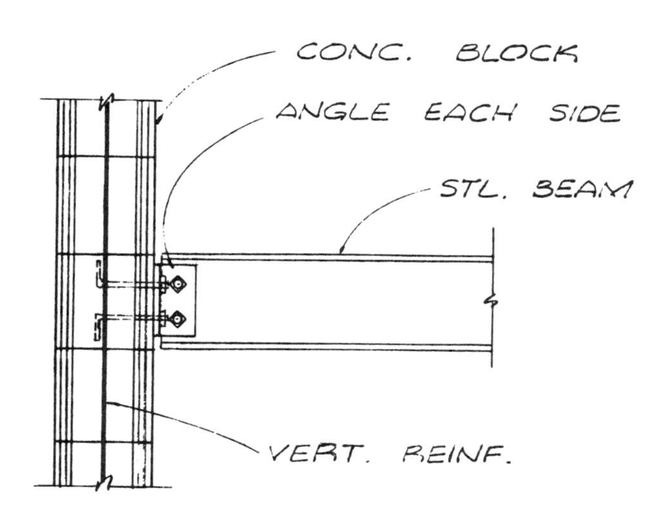

Detail 23. A section of a concrete block masonry wall supporting a steel beam. The steel beam reaction is transferred to the wall through clip angles bolted to each side of the beam web and into the masonry wall. The size of the angles and the size of the bolts are determined by calculation.

Notes ▪ Drawings ▪ Ideas

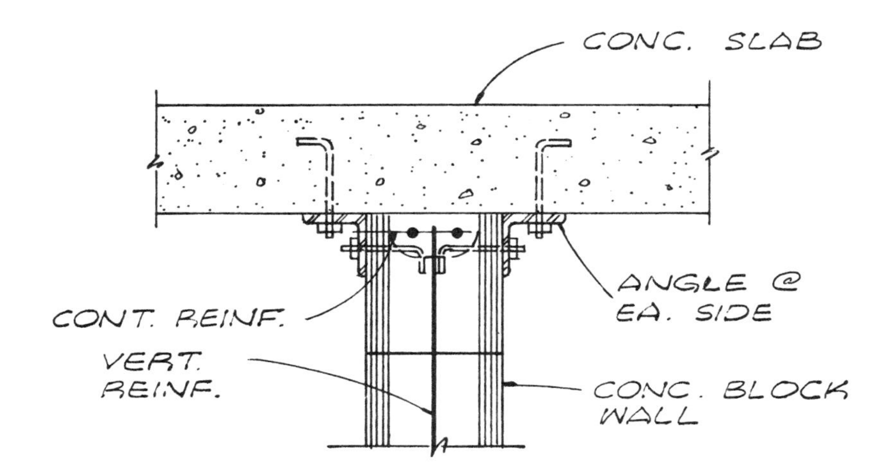

Detail 24(a). A section of the top of a concrete block masonry wall connected to a poured concrete slab. The wall is connected to the slab by continuous angles on each side of the wall. The angles are bolted to the slab and to the wall as shown. The size and spacing of the angle bolts are determined by calculation.

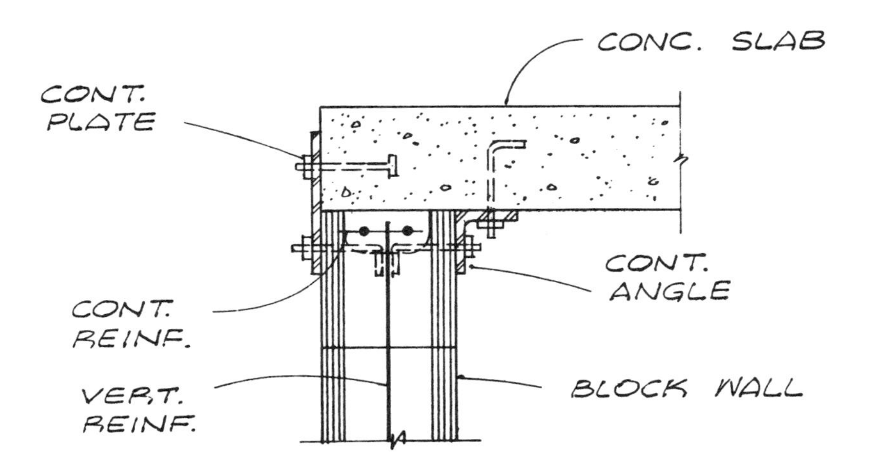

Detail 24(b). A section of the top of a concrete block masonry wall connected to the edge of a poured concrete slab. The wall is connected to the slab by a continuous angle and a continuous flat plate. The angle and plate are bolted to the slab and to the wall as shown. The size and spacing of the bolts are determined by calculation.

Notes ▪ Drawings ▪ Ideas

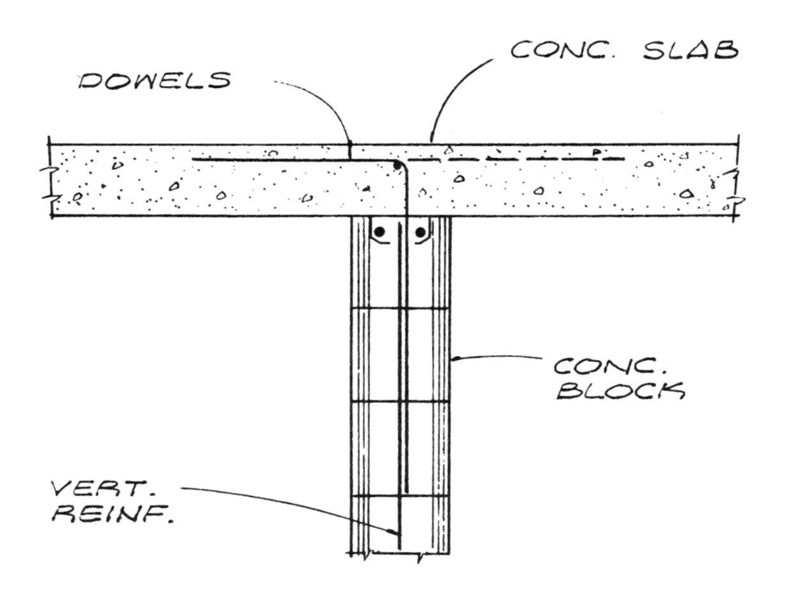

Detail 25(a). A section of a concrete slab supported by a concrete block masonry wall. The slab is connected to the wall by reinforcing dowels bent in alternate directions as shown. Continuous horizontal reinforcing bars are added at the top of the wall and at the top of the slab. The dowels lap 30 bar diameters in the masonry and 40 bar diameters in the concrete or not less than 24".

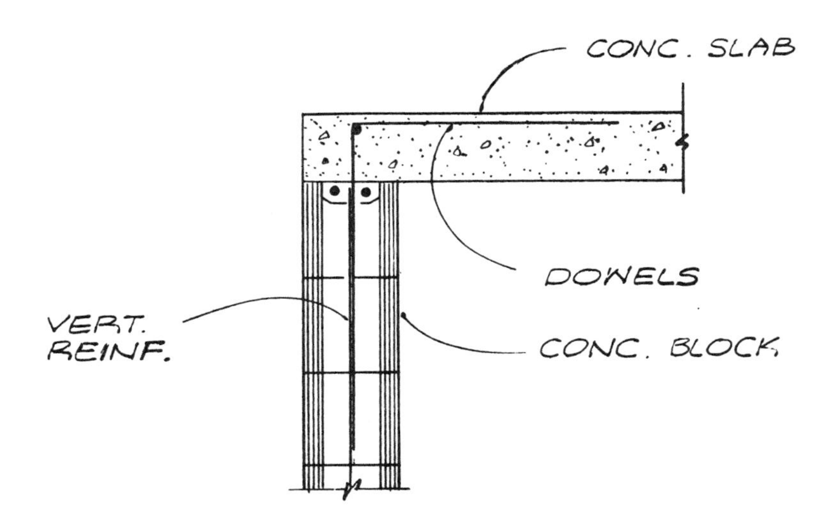

Detail 25(b). A section of a concrete slab supported by a concrete block masonry wall. The slab is connected to the wall by reinforcing dowels as shown. Continuous horizontal reinforcing bars are added at the top of the wall and at the top of the slab. The dowels lap 30 bar diameters in the masonry and 40 bar diameters in the concrete or not less than 24".

Notes ▪ Drawings ▪ Ideas

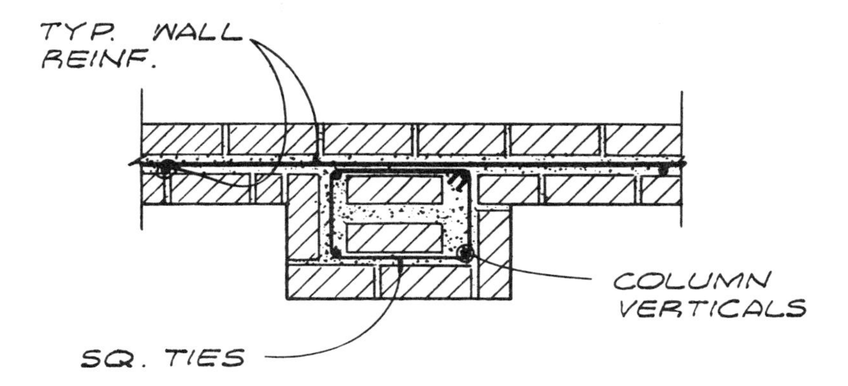

Detail 26(a). A plan section of a brick wall rectangular pilaster. The pilaster is reinforced with a minimum of four vertical bars tied together in the same way as is required for a rectangular concrete column. The interior bricks shown in the pilaster plan may be omitted and the total section filled with grout or concrete.

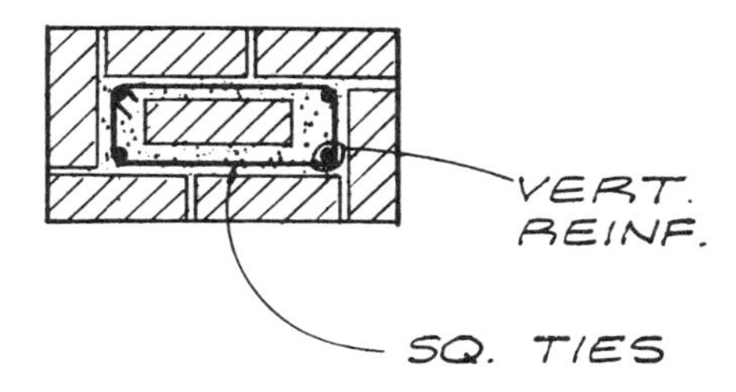

Detail 26(b). A section of a rectangular brick column. The column is reinforced with a minimum of four vertical bars tied together in the same way as is required for a rectangular concrete column. The interior brick shown in the column plan may be omitted and the total section filled with grout or concrete.

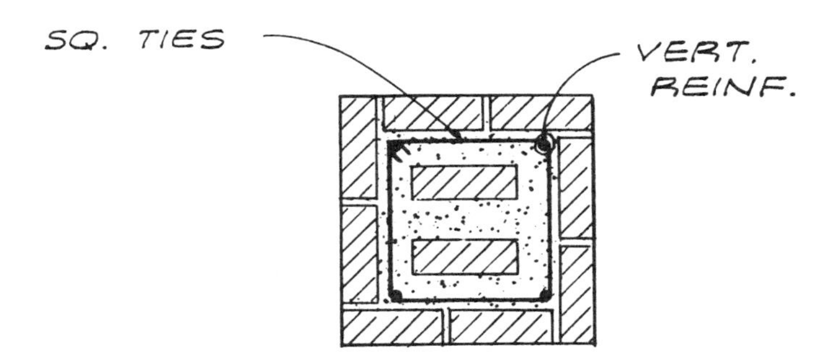

Detail 26(c). A section of a square brick column. The column is reinforced with a minimum of four vertical bars tied together in the same way as is required for a rectangular concrete column. The interior bricks shown in the column plan may be omitted and the total section filled with grout or concrete.

Notes ▪ Drawings ▪ Ideas

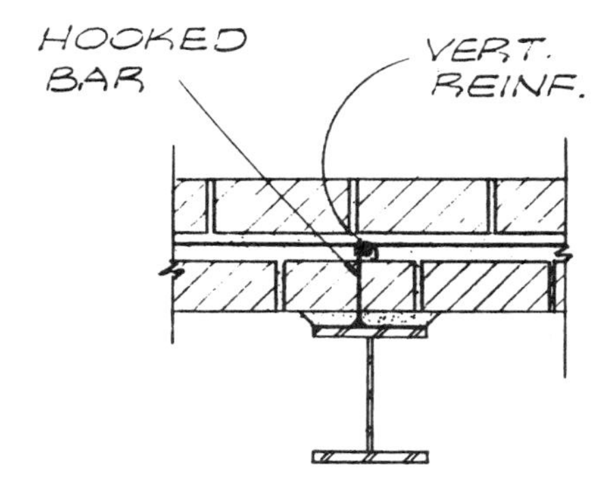

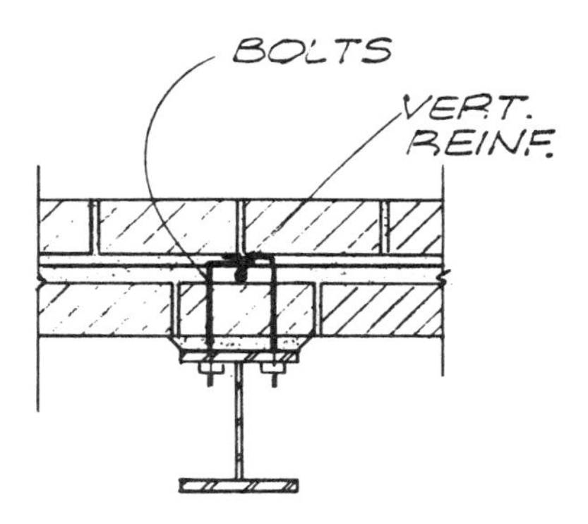

Detail 27(a). A plan section of a brick wall and a wide flange steel column. The column is connected to the wall by reinforcing bars bent as hooks, welded to the flange of the column, and hooked around a vertical reinforcing bar in the wall. The wall restrains the column laterally. The space between the face of the wall and the face of the column flange is filled with grout.

Detail 27(b). A plan section of a brick wall and a wide flange steel column. The column is bolted to the wall as shown. The wall restrains the column laterally. A vertical reinforcing bar is added in the wall at the line of the column connection. The space between the face of the wall and the face of the column flange is filled with grout.

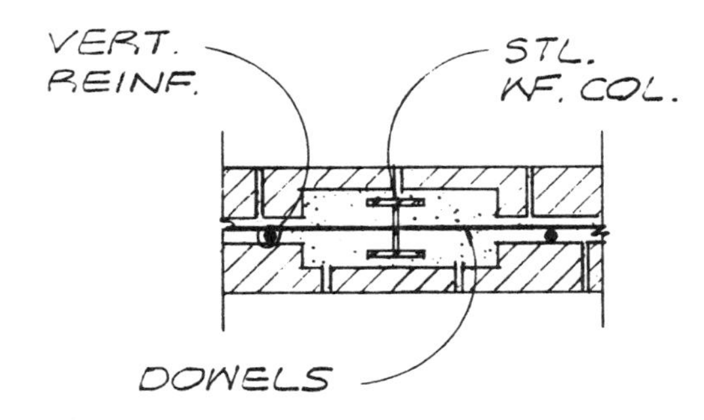

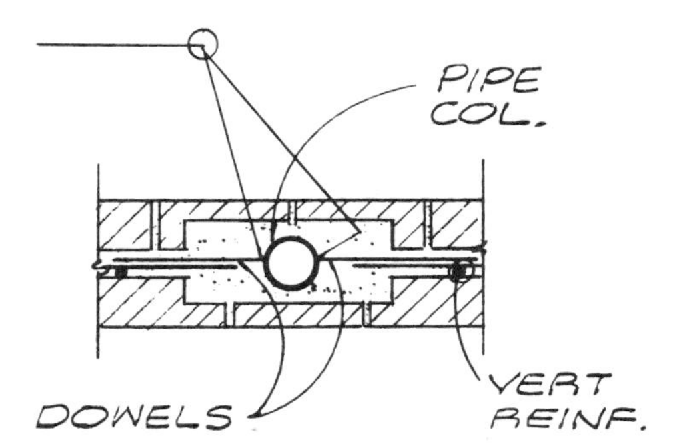

Detail 27(c). A plan section of a wide flange steel column inside a brick wall. Horizontal reinforcing dowels are welded to each side of the column web or pass through a hole in the web and extend 24″ into the wall grout space. The horizontal dowels are spaced at 4′0″ o.c.

Detail 27(d). A plan section of a steel pipe column inside a brick wall. Horizontal reinforcing dowels are welded to the column on each side and extend 24″ into the wall grout space. The horizontal dowels are spaced at 4′0″ o.c.

Notes ▪ Drawings ▪ Ideas

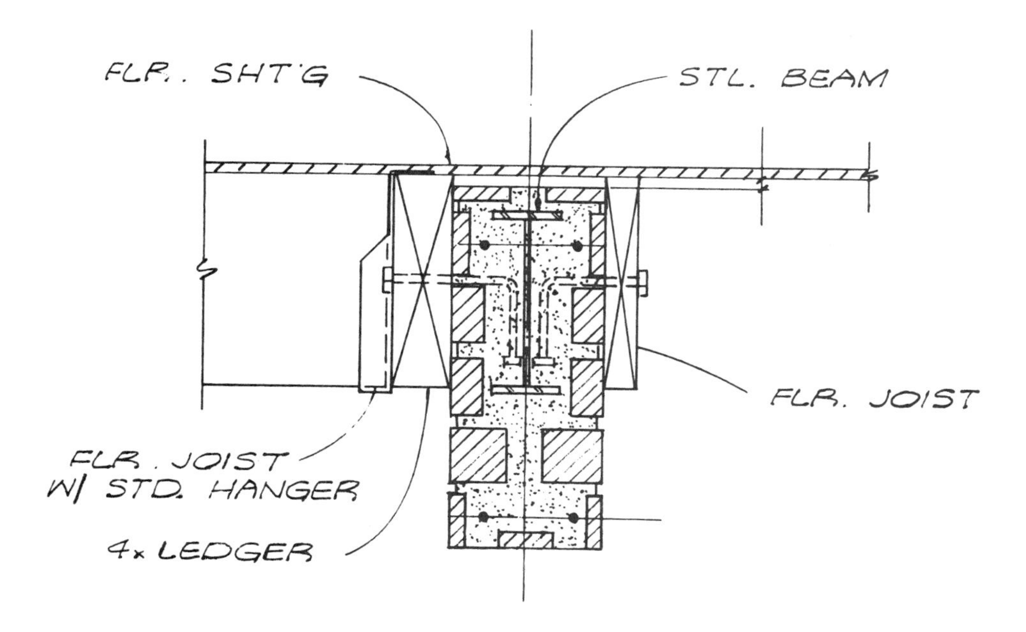

Detail 28(a). A section of a composite brick and steel beam. The floors are connected to the beam as shown in Detail 36 and similar to Detail 39. A space is provided between the top of the brick beam and the bottom of the sheathing to allow for wood shrinkage.

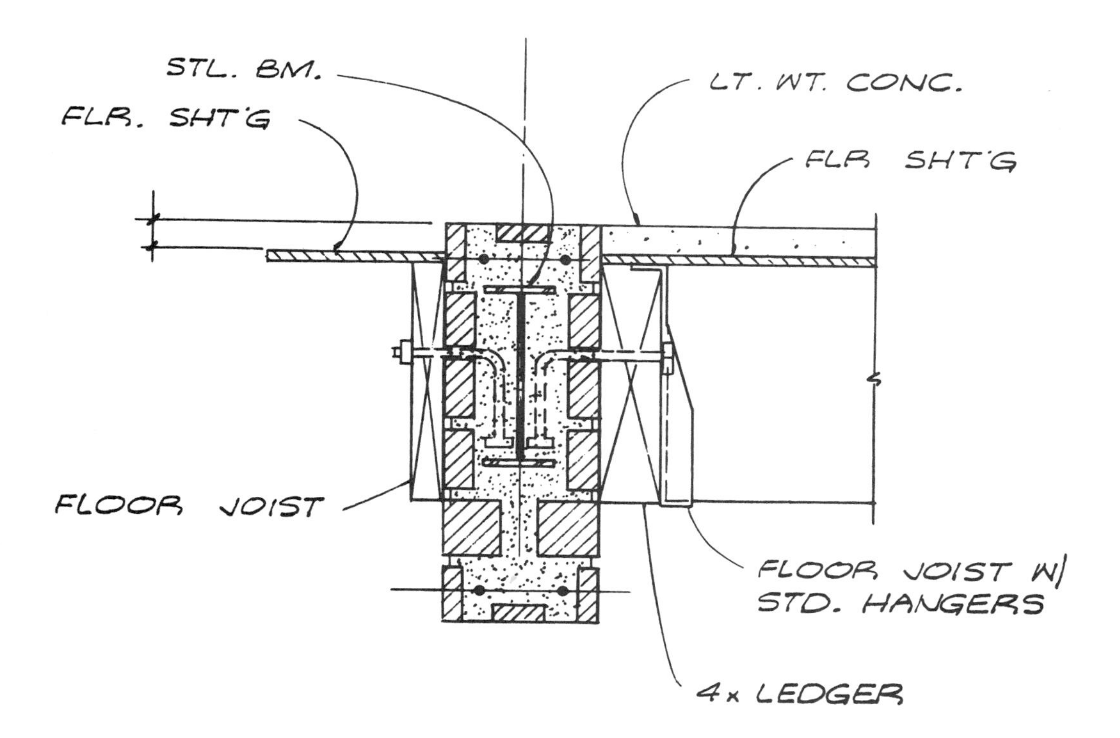

Detail 28(b). A section of a composite brick and steel beam. The floor joists are connected to the beam as shown in Detail 36 and similar to Detail 39. The exterior wood floor is depressed below the top of the brick beam and the interior finished floor.

Notes ▪ Drawings ▪ Ideas

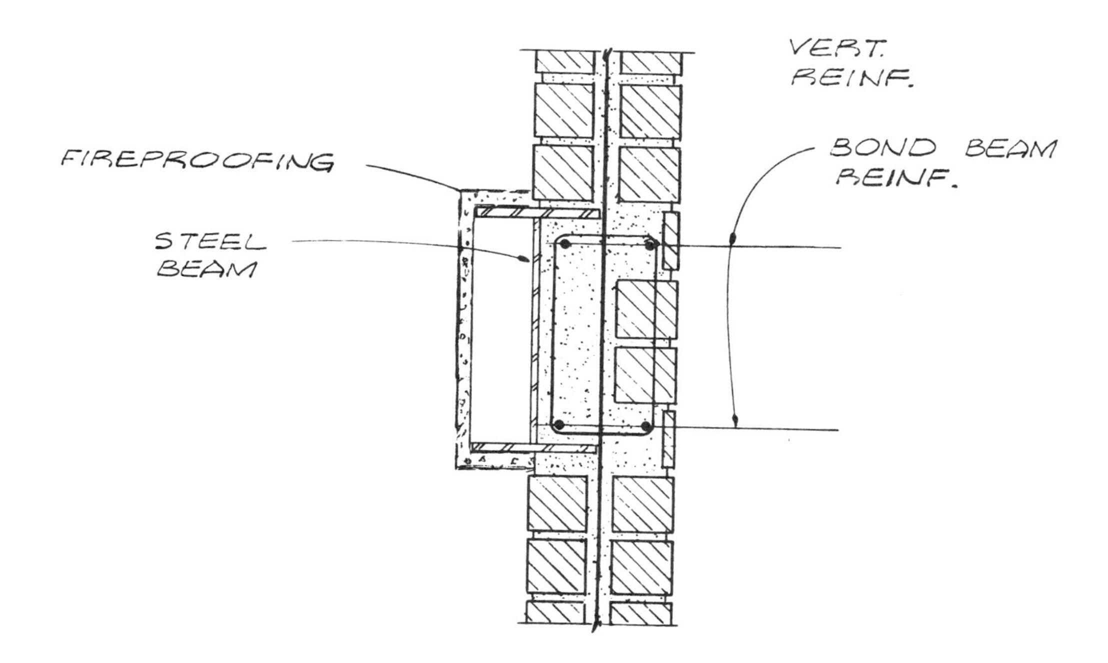

Detail 29. A section of a brick wall and a steel beam. The exterior surface of the steel beam is covered with a fireproof material. The inside face of the beam web acts as a form for the bond beam in the brick wall. The flange of the steel beam should not interrupt the vertical reinforcing of the brick wall; however point contact is permitted.

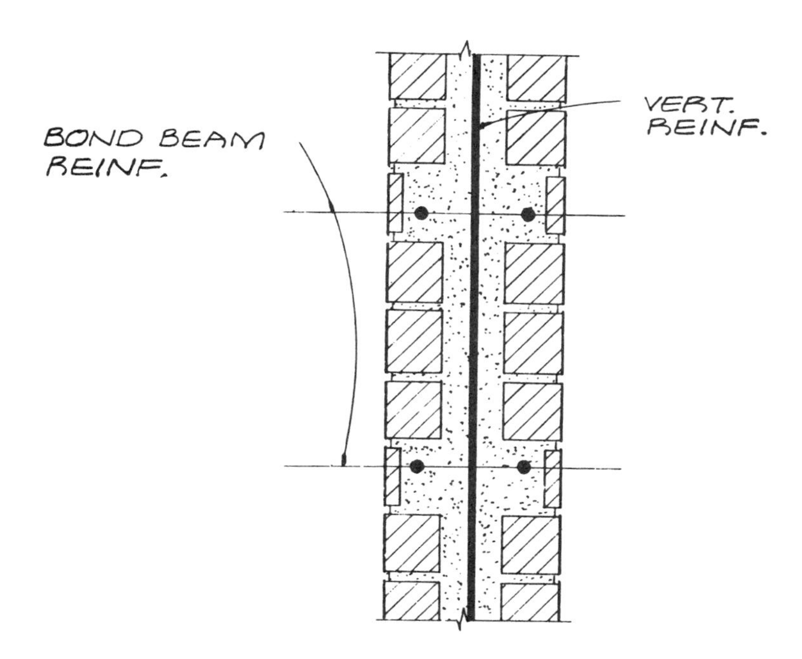

Detail 30. A section of a brick wall bond beam. The vertical distance between the horizontal reinforcement depends on the required area of the bond beam. The size of the horizontal reinforcement is determined by the load to be transferred through the masonry wall. The split bricks provide grout coverage of the reinforcing steel.

Notes ▪ Drawings ▪ Ideas

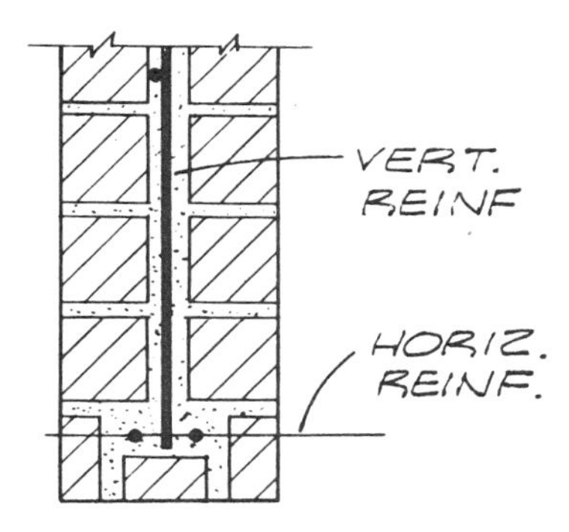

Detail 31(a). A section of a brick wall lintel or sill. The vertical reinforcement is stopped at the edge of the wall opening. The surface of the wall edge is covered by a split brick. The horizontal reinforcing bars extend 24" past the edge of the opening.

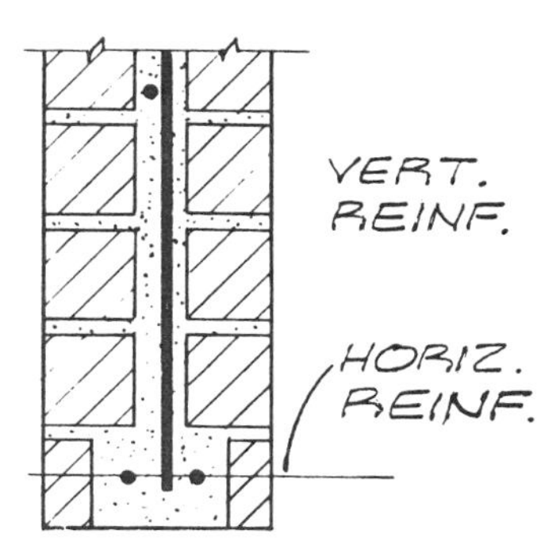

Detail 31(b). A section of a brick wall lintel or sill. The vertical reinforcement is stopped at the edge of the wall opening. The surface of the wall edge is covered by the grout fill. The horizontal reinforcing bars extend 24" past the edge of the opening.

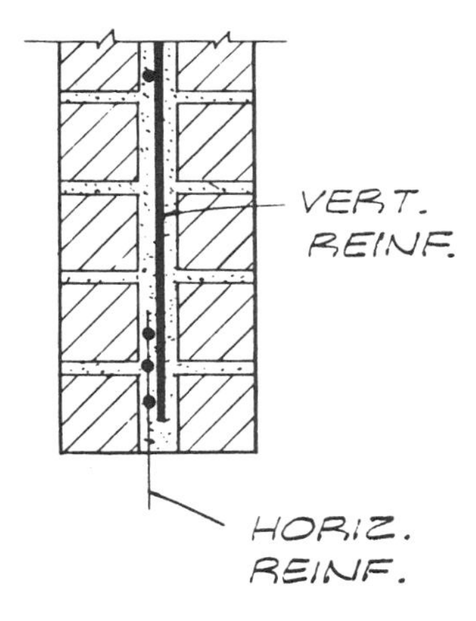

Detail 32(a). A section of a brick wall lintel or sill. The vertical reinforcement is stopped at the edge of the wall opening. The horizontal reinforcement is placed as shown. The horizontal reinforcing bars extend 24" past the edge of the opening.

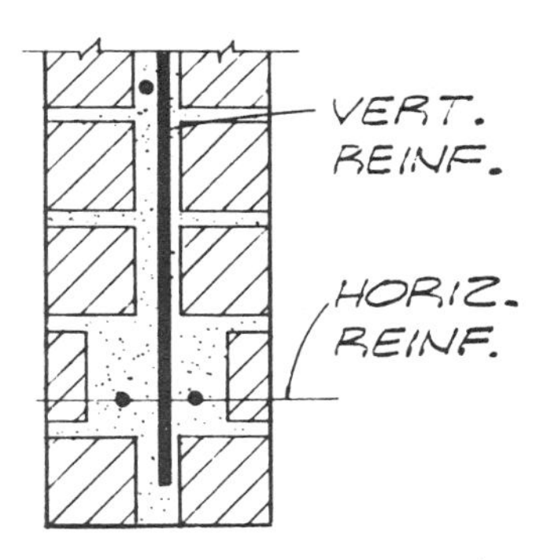

Detail 32(b). A section of a brick wall lintel or sill. The vertical reinforcement is stopped at the edge of the wall opening. The split bricks provide minimum grout coverage over the reinforcing steel. The horizontal reinforcing bars extend 24" past the edge of the opening.

Notes ▪ Drawings ▪ Ideas

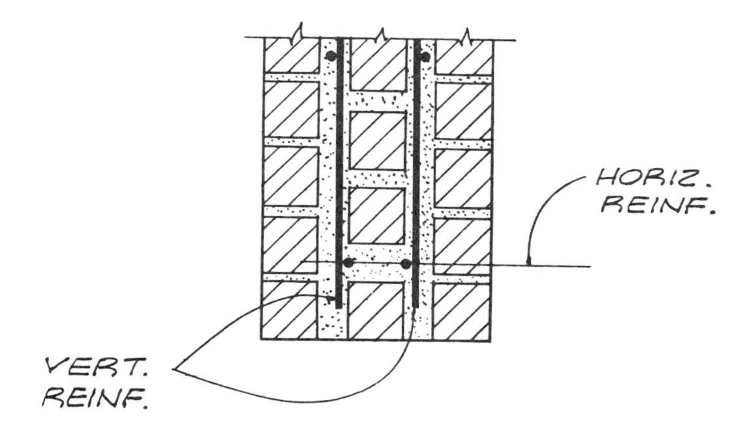

Detail 33(a). A section of a brick wall lintel or sill. The wall is constructed of three tiers of brick. The vertical reinforcement is stopped at the edge of the wall opening. The horizontal reinforcing bars extend 24" past the edge of the opening.

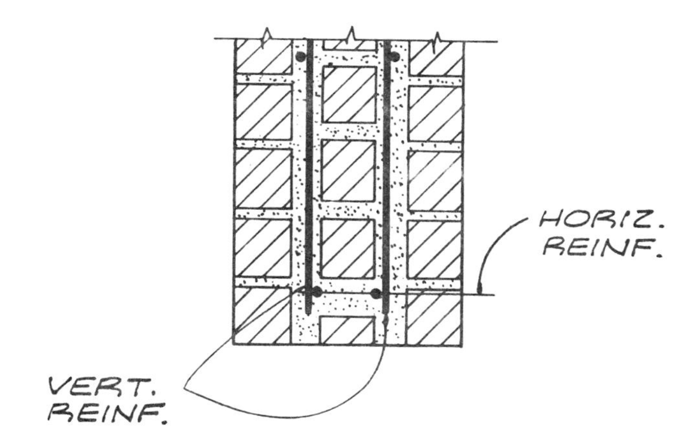

Detail 33(b). A section of a brick wall lintel or sill. The wall is constructed of three tiers of brick. The vertical reinforcement is stopped at the edge of the wall opening. A split brick is used at the surface of the edge of the wall to provide minimum grout coverage of the reinforcing steel. The horizontal reinforcing bars extend 24" past the edge of the opening.

Notes ▪ Drawings ▪ Ideas

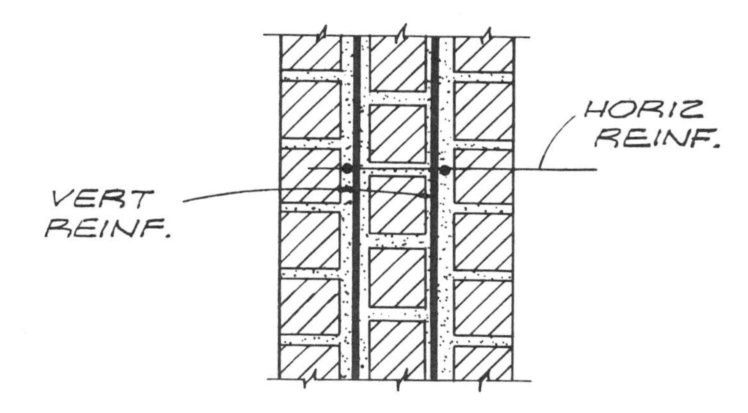

Detail 34(a). A typical section of a brick masonry wall three tiers wide and with two rows of horizontal and vertical reinforcement. See Detail 35(b) for reinforcing bar clearance and grout coverage in masonry walls; see also Detail 34(b).

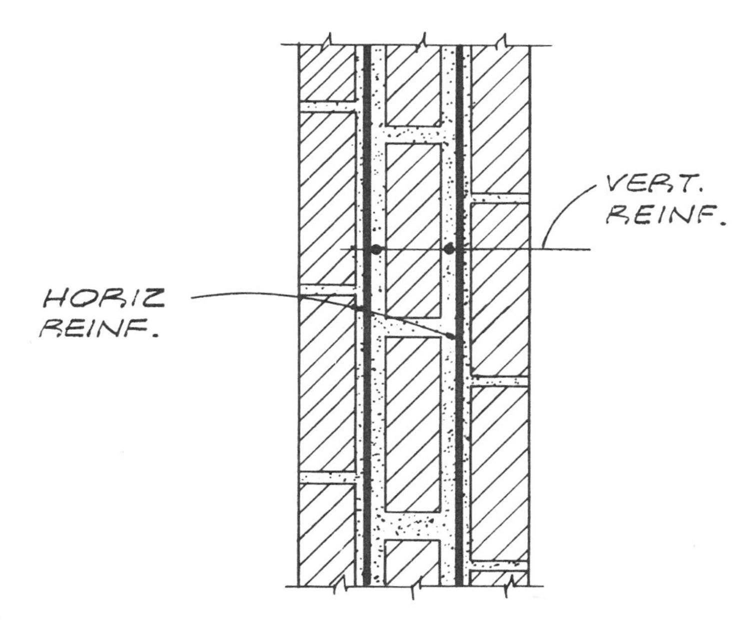

Detail 34(b). A plan section of a brick wall three tiers wide with two rows of horizontal and vertical reinforcement. See Detail 34(a).

Notes ▪ Drawings ▪ Ideas

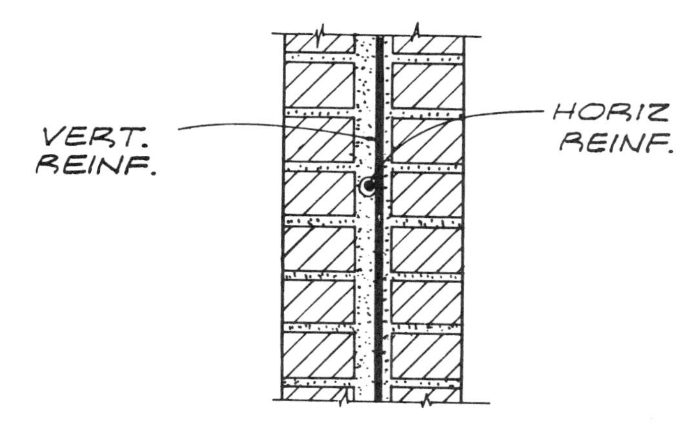

Detail 35(a). A typical section of a brick wall two tiers wide with one row of horizontal and vertical reinforcement. See Detail 35(b) for reinforcing bar clearance and grout coverage in masonry walls.

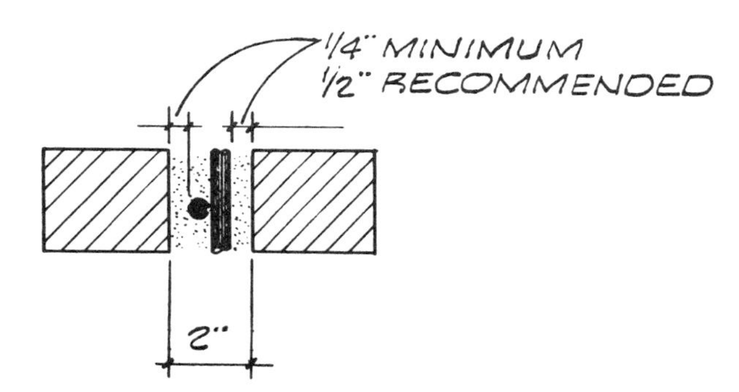

Detail 35(b). A section of brick masonry and reinforcing bars showing the required clearance between the brick surface and the reinforcing bars. The horizontal and vertical reinforcing bars are permitted point contact. It is recommended that the total grout space be not less than 2″.

Notes ▪ Drawings ▪ Ideas

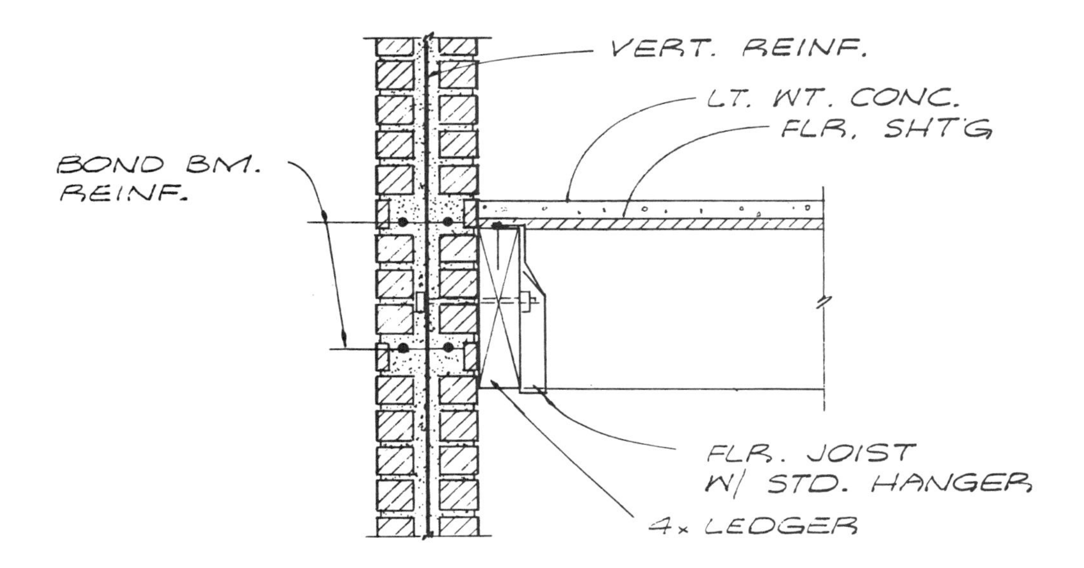

Detail 36. A section of a brick wall supporting floor joists and light weight concrete over the wood floor. The floor joists are connected to a 4" wide wood ledger by standard joist hangers. The ledger is bolted to the wall. The size and spacing of the ledger bolts are determined by calculation. The floor sheathing is nailed to the ledger to transfer floor diaphragm stress into the wall. The horizontal reinforcing bars act as a bond beam in the masonry wall.

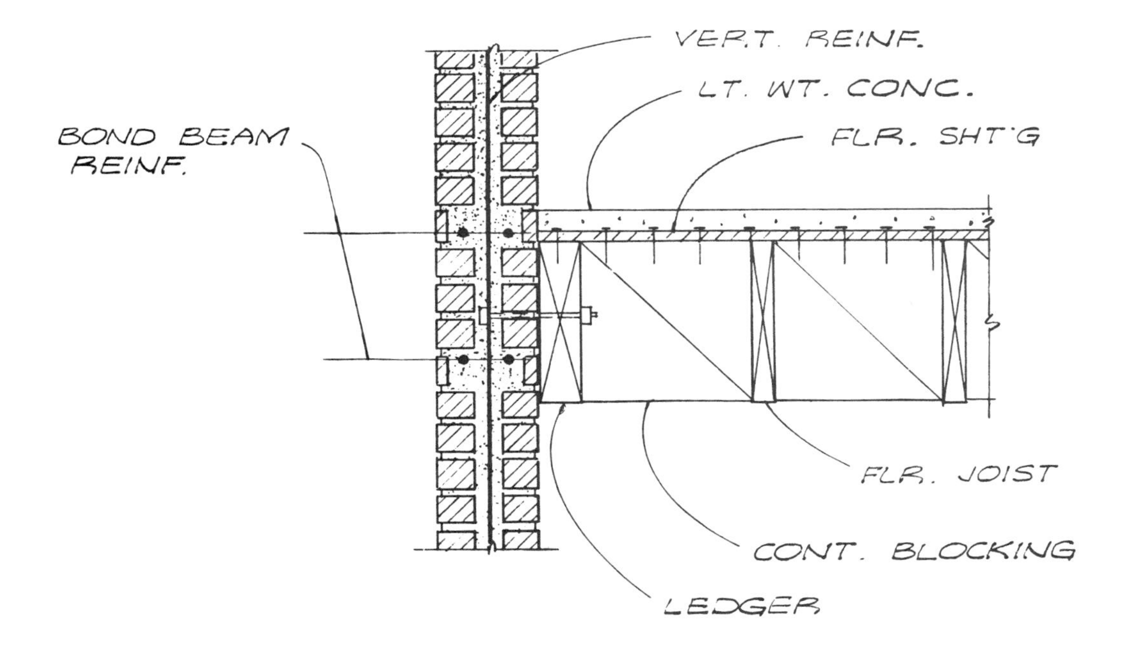

Detail 37. A section of a brick wall, floor joists, and light-weight concrete over the wood floor. The floor joists span parallel to the wall. The floor sheathing is nailed to a 4" wide wood ledger to transfer the floor diaphragm stress into the wall. The diaphragm stress may require that the perimeter floor joists be blocked and nailed as shown. The ledger is bolted to the wall. The size and spacing of the ledger bolts are determined by calculation. The horizontal reinforcing bars act as a bond team in the wall.

Notes ▪ Drawings ▪ Ideas

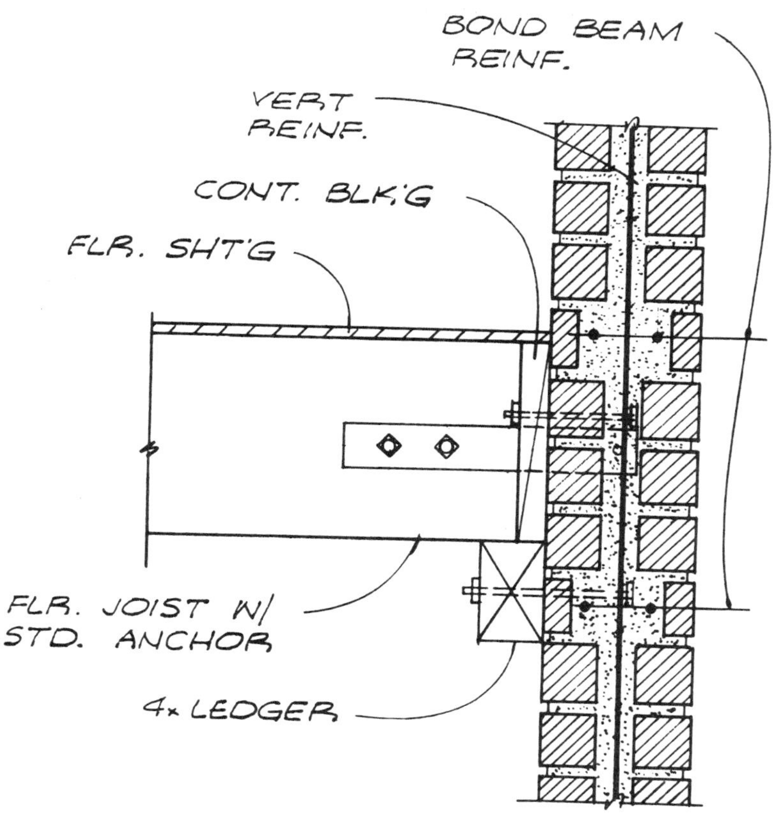

Detail 38. A section of a brick wall, floor joists, and light-weight concrete over a wood floor. The joists bear on a 4" wide wood ledger which is bolted to the wall. The size and spacing of the ledger bolts are determined by calculation. The joist continuous blocking is bolted to the wall.

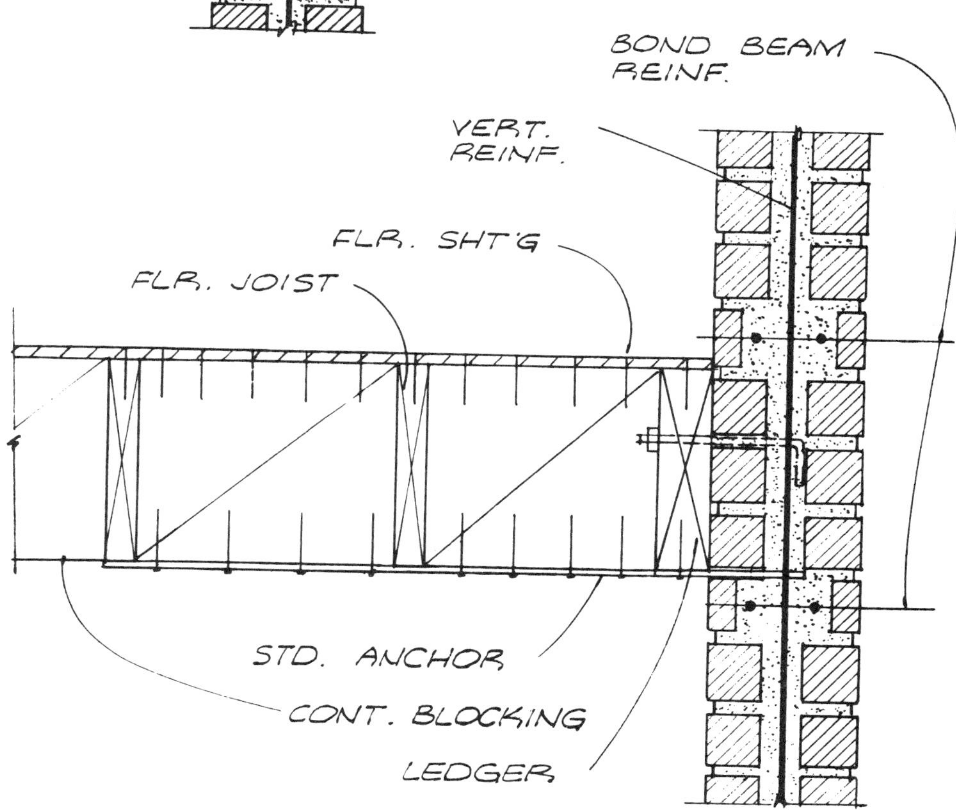

Detail 39. A section of a brick wall, floor joists, and floor sheathing. The floor joists span parallel to the wall. The floor sheathing is nailed to a 4" wide ledger to transfer the diaphragm stress into the wall. The diaphragm stress may require that the perimeter floor joists be blocked and nailed as shown. The ledger is bolted to the masonry wall. The size and spacing of the ledger bolts are determined by calculation. The horizontal reinforcing bars act as a bond beam in the wall. A metal tie strap spaced at 4'0" connects the floor to the wall.

Notes ▪ Drawings ▪ Ideas

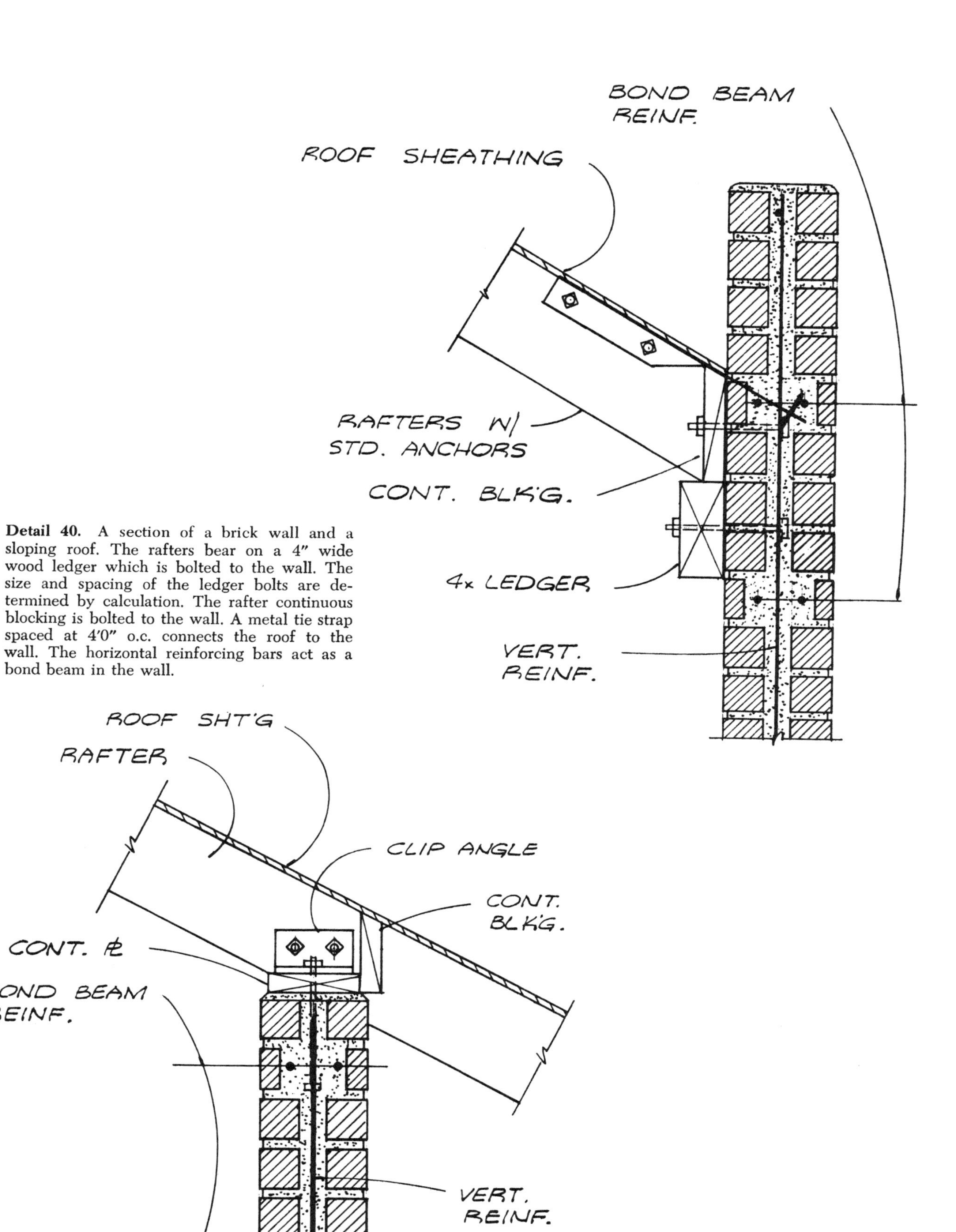

Detail 40. A section of a brick wall and a sloping roof. The rafters bear on a 4" wide wood ledger which is bolted to the wall. The size and spacing of the ledger bolts are determined by calculation. The rafter continuous blocking is bolted to the wall. A metal tie strap spaced at 4'0" o.c. connects the roof to the wall. The horizontal reinforcing bars act as a bond beam in the wall.

Detail 41. A section of a brick wall supporting a sloping roof. The rafters are connected to a sill plate at the top of the wall by a clip angle spaced at 4'0" o.c. The horizontal reinforcing bars act as a bond beam in the masonry wall.

Notes ▪ Drawings ▪ Ideas

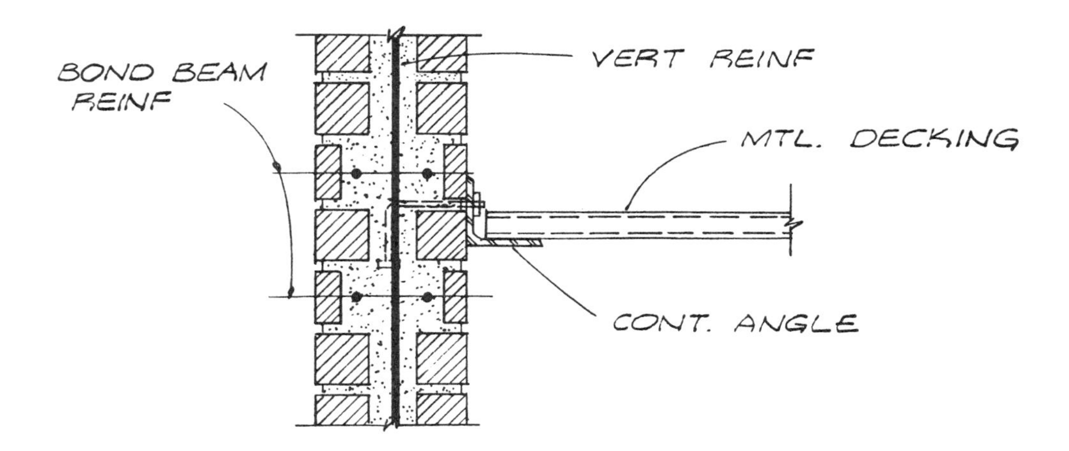

Detail 42. A section of a brick wall and exposed metal decking. The decking is supported by and welded to a continuous ledger angle. The ledger angle is bolted to the wall. The size and spacing of the angle and bolts are determined by calculation. The horizontal reinforcing bars act as a bond beam in the wall.

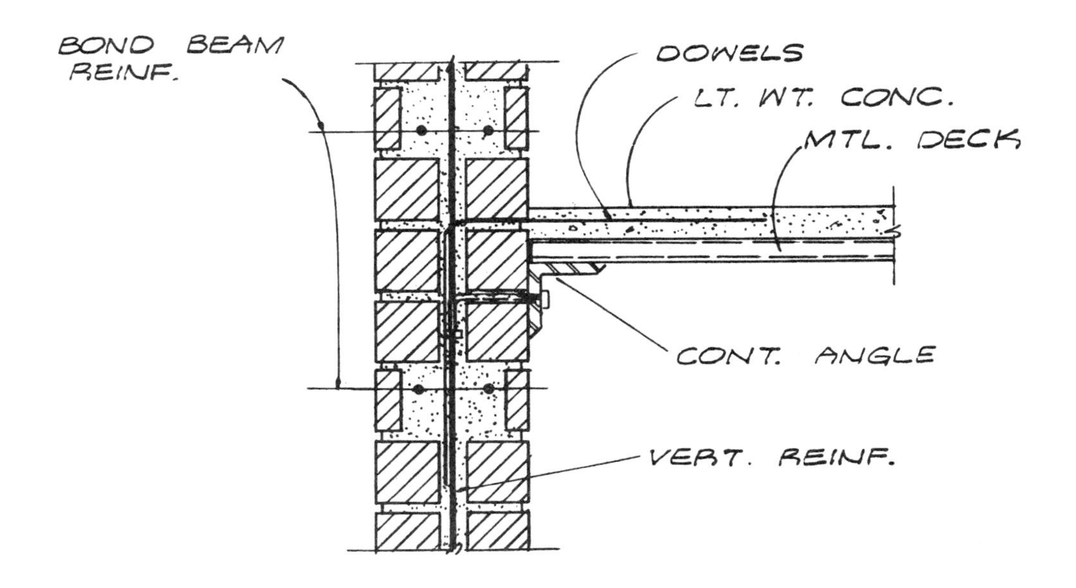

Detail 43. A section of a brick wall and metal decking with lightweight concrete. The decking is supported by a continuous ledger angle which is bolted to the wall. The size and spacing of the angle and bolts are determined by calculation. The slab diaphragm stress is transferred to the wall by the reinforcing dowels. The horizontal bars act as a bond beam in the wall.

Notes ▪ Drawings ▪ Ideas

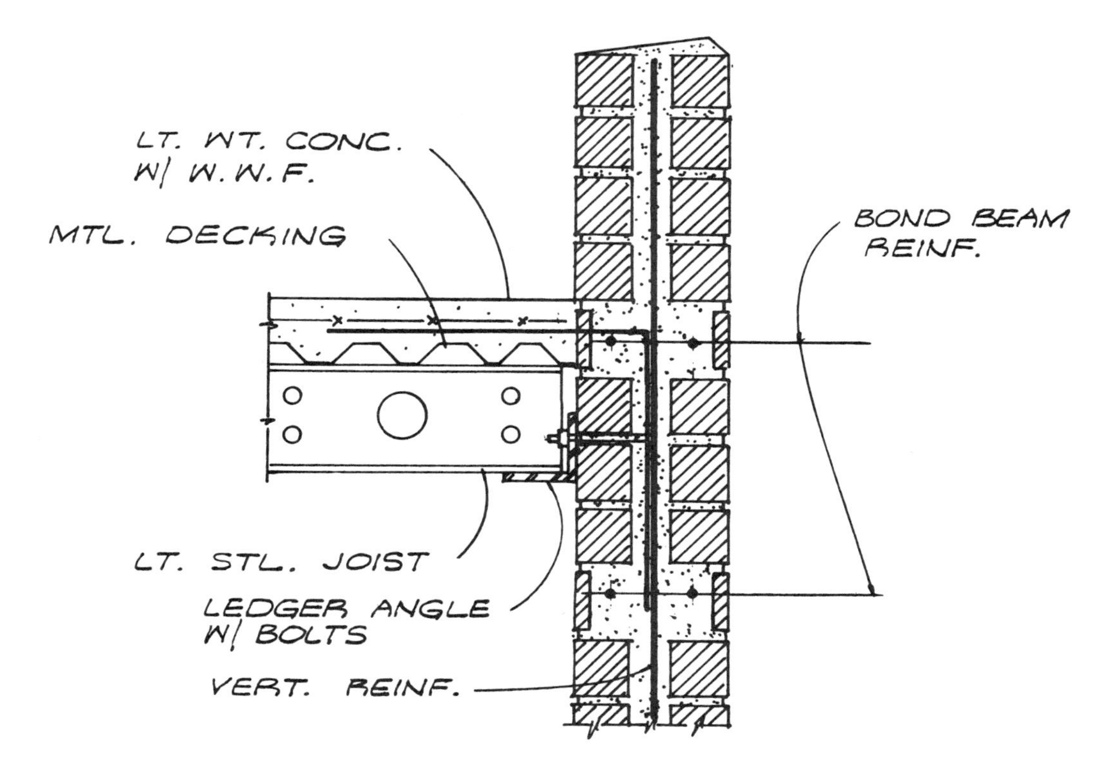

Detail 44. A section of a brick wall, light steel joists, a continuous ledger angle, and metal decking with light-weight concrete. The steel joists bear on and are welded to the ledger angle which is bolted to the wall. The size and spacing of the angle and bolts are determined by calculation. The light-weight concrete slab is connected to the wall with reinforcing dowels. The horizontal reinforcing bars act as a bond beam in the wall.

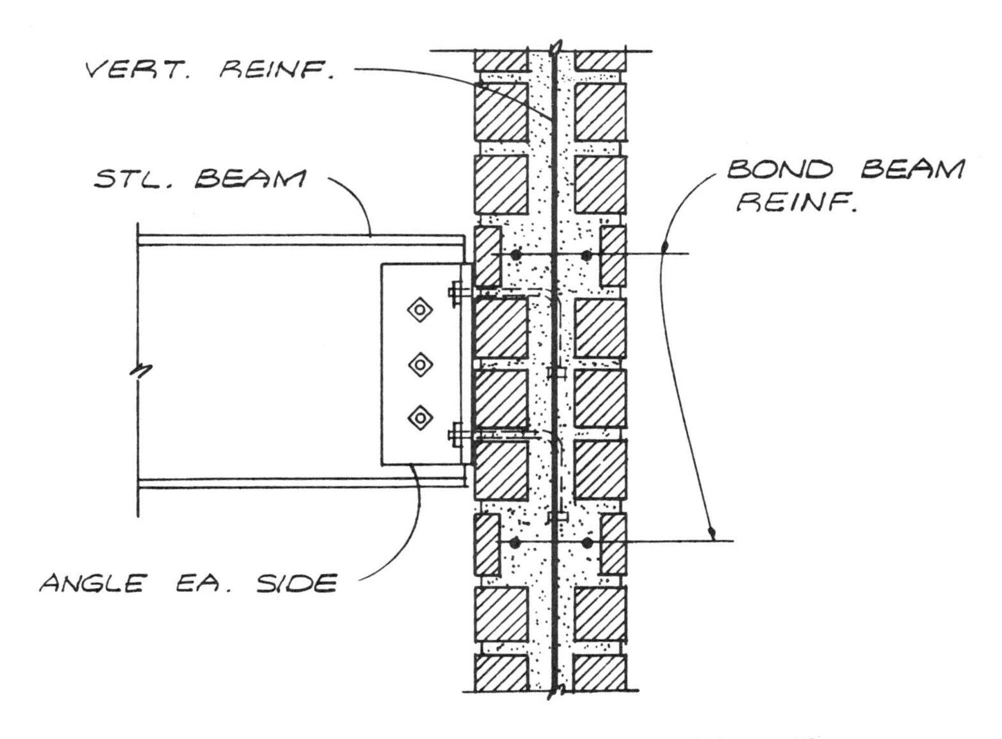

Detail 45. A section of a brick wall supporting a steel beam. The steel beam reaction is transferred to the wall through clip angles bolted to each side of the beam web and into the wall. The size of the angles and the bolts are determined by calculation.

Notes ▪ Drawings ▪ Ideas

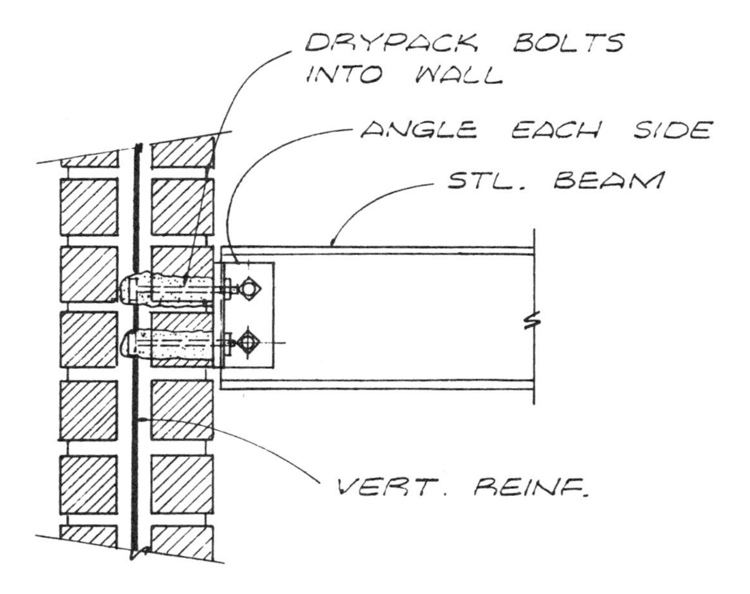

Detail 46. A section of a brick wall supporting a steel beam. The steel beam reaction is transferred to the wall through clip angles bolted to each side of the beam web and into the wall. The size of the angles and the bolts are determined by calculation. This detail is used to connect steel beams to existing walls.

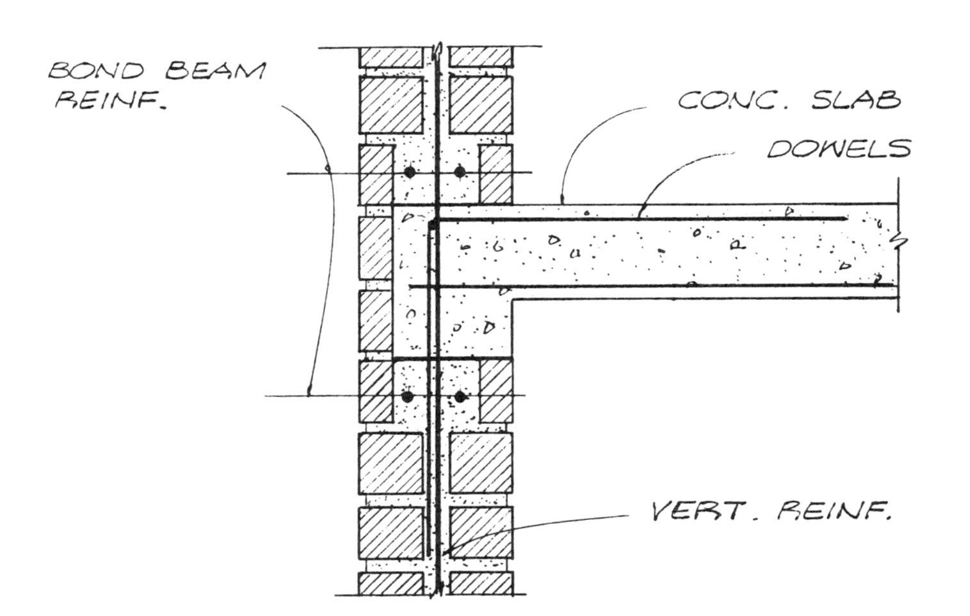

Detail 47. A section of a brick wall and a poured concrete slab. The slab is cast into the wall for vertical support. The slab diaphragm stress is transferred to the wall by the reinforcing dowels.

Notes ▪ Drawings ▪ Ideas

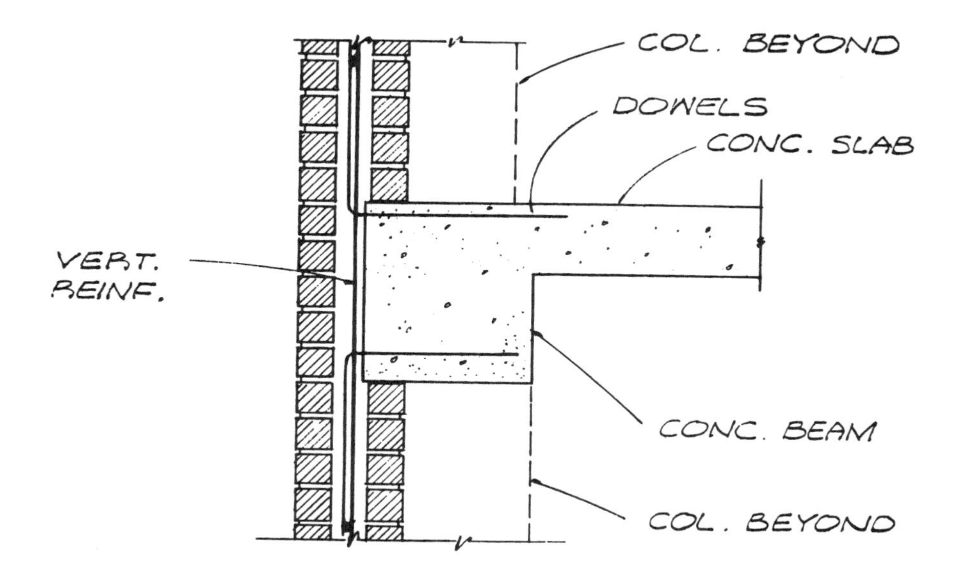

Detail 48. A section of a brick wall, a concrete slab, and a concrete spandrel beam. The spandrel beam is cast into the wall as shown. The wall is tied to the beam with reinforcing dowels. The beam and slab reinforcement is not shown.

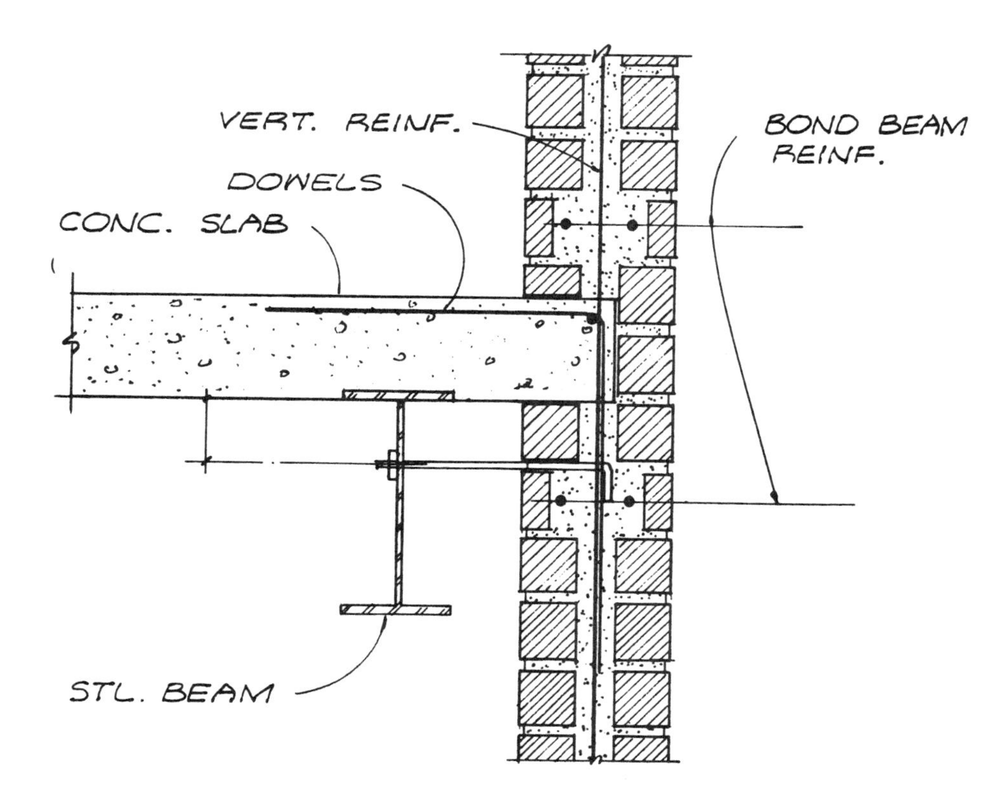

Detail 49. A section of a brick wall and a concrete slab. The slab is supported by a steel beam. The slab diaphragm stress is transferred to the wall by reinforcing dowels. Equally spaced rods between the web of the steel beam and the brick wall restrain the beam laterally. The horizontal reinforcing bars act as a bond beam in the wall.

Notes ▪ Drawings ▪ Ideas

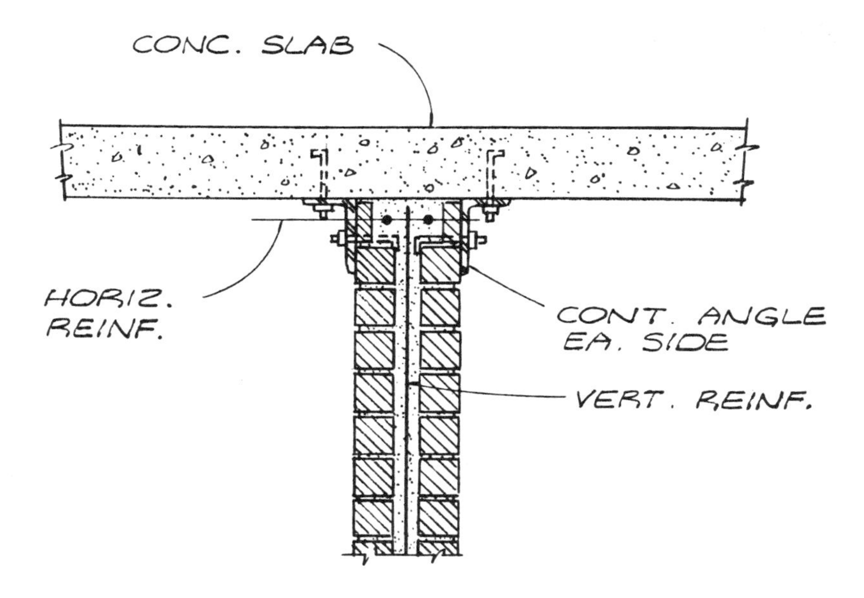

Detail 50(a). A section of the top of a brick wall connected to a poured concrete slab. The wall is connected to the slab by continuous angles on each side of the wall. The angles are bolted to the slab and to the wall as shown. The size and spacing of the bolts are determined by calculation.

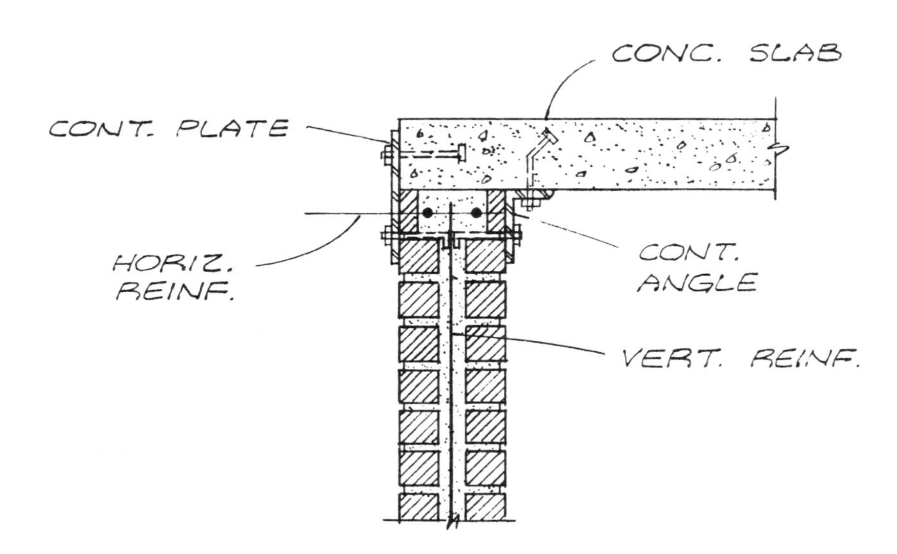

Detail 50(b). A section of the top of a brick wall connected to the edge of a poured concrete slab. The wall is connected to the slab by a continuous angle and a flat plate. The angle and plate are bolted to the slab and to the wall as shown. The size and spacing of the bolts are determined by calculation.

Notes ▪ Drawings ▪ Ideas

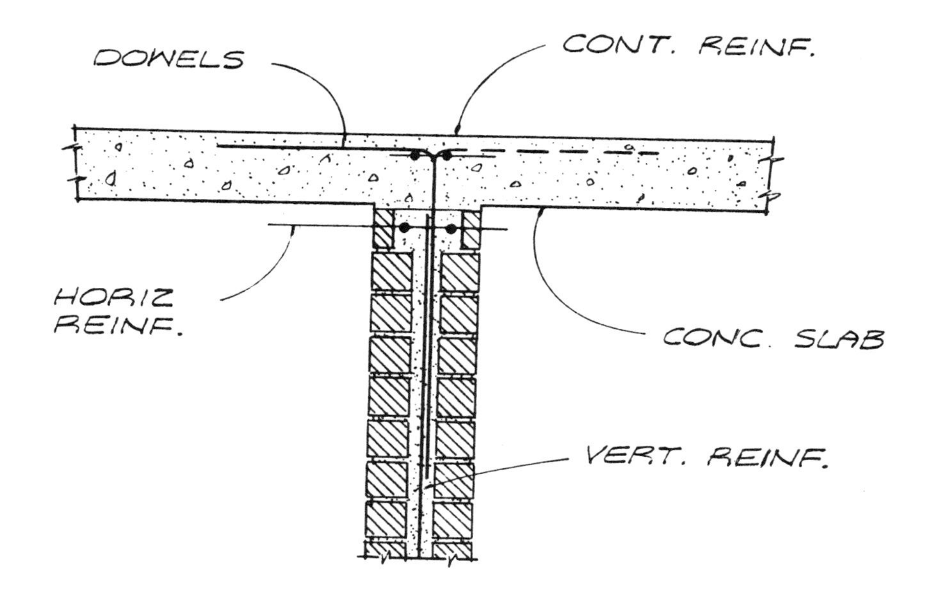

Detail 51(a). A section of a concrete slab supported by a brick wall. The slab is connected to the wall by reinforcing dowels bent in alternate directions as shown. Two horizontal reinforcing bars are added at the top of the wall and at the top of the slab. The dowels lap 30 bar diameters in the masonry and 40 bar diameters in the concrete or not less than 24".

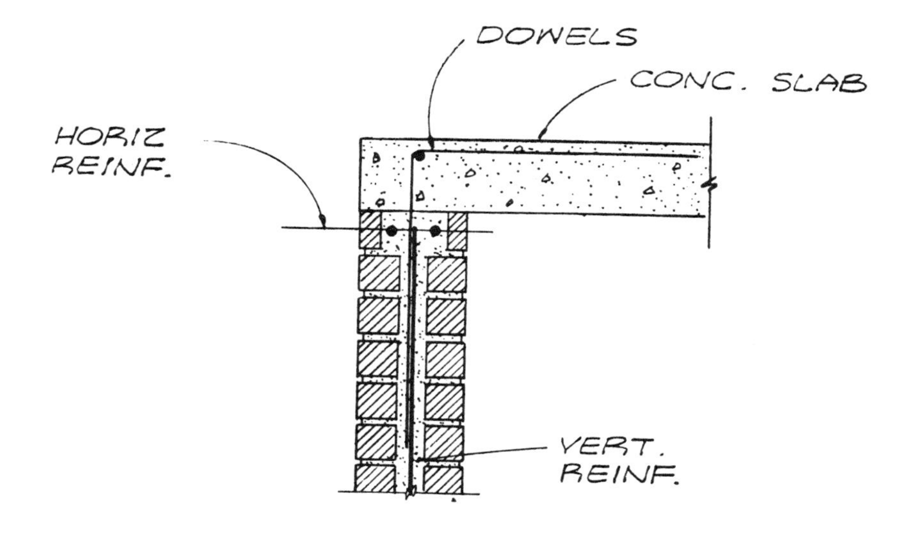

Detail 51(b). A section of a concrete slab supported by a brick wall. The slab is connected to the wall by reinforcing dowels as shown. Continuous horizontal reinforcing bars are added at the top of the wall and at the top of the slab. The dowels lap 30 bar diameters in the masonry and 40 bar diameters in the concrete or not less than 24".

Notes ▪ Drawings ▪ Ideas

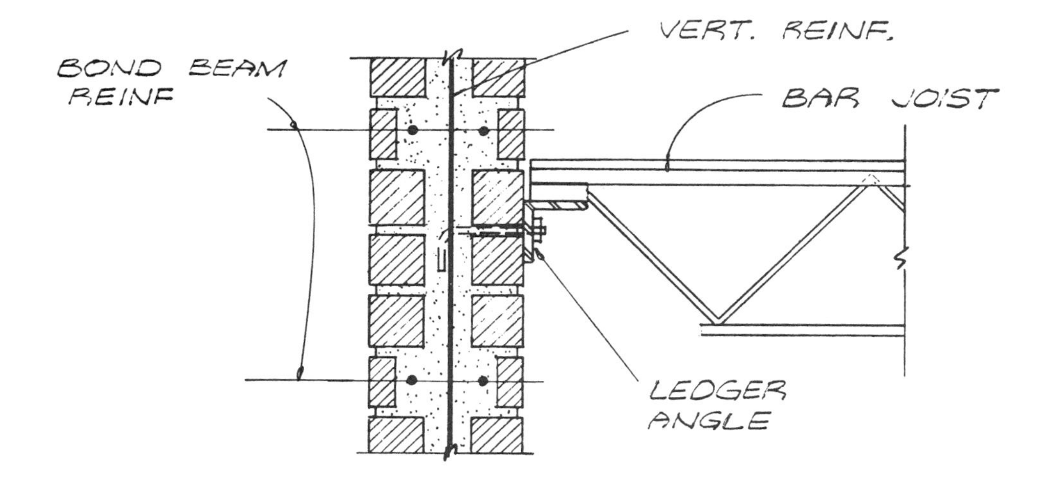

Detail 52. A section of a brick wall, metal decking, and steel bar joists. The bar joists are supported by and welded to a continuous ledger angle. The ledger angle is bolted to the brick wall. The size and spacing of the ledger bolts are determined by calculation. The concrete diaphragm stress is transferred to the wall by reinforcing dowels. The horizontal bars act as a bond beam in the wall.

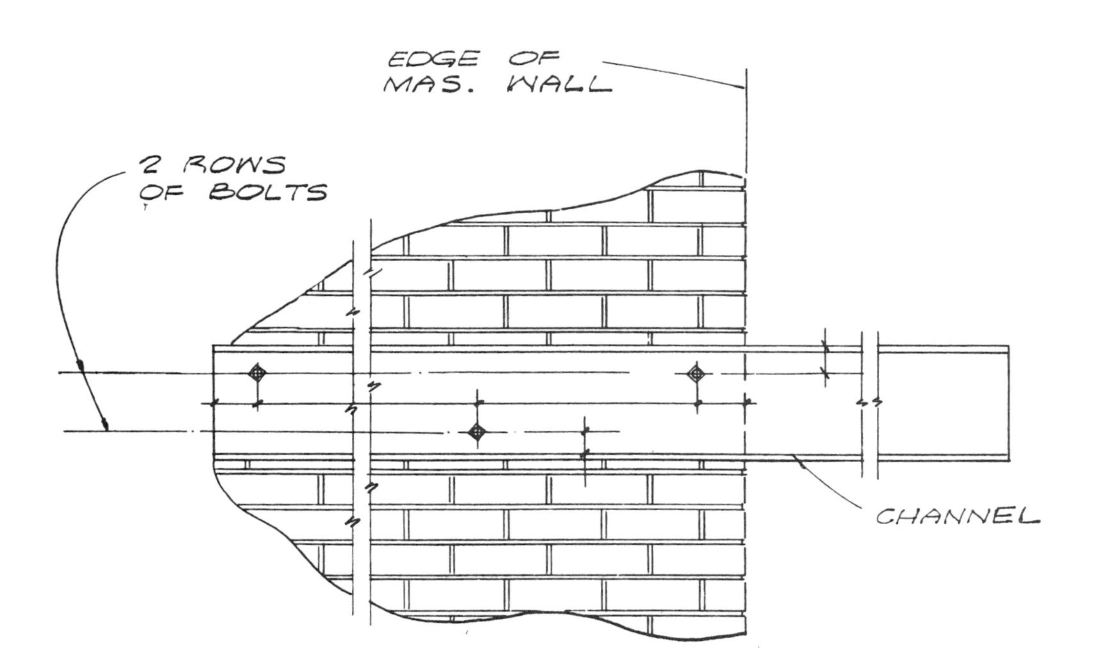

Detail 53. An elevation of a steel channel bolted to and cantilevered from the face of a masonry wall. The channel web is bolted to the wall as shown. The size and spacing of the bolts are determined by calculation.

Notes ▪ Drawings ▪ Ideas

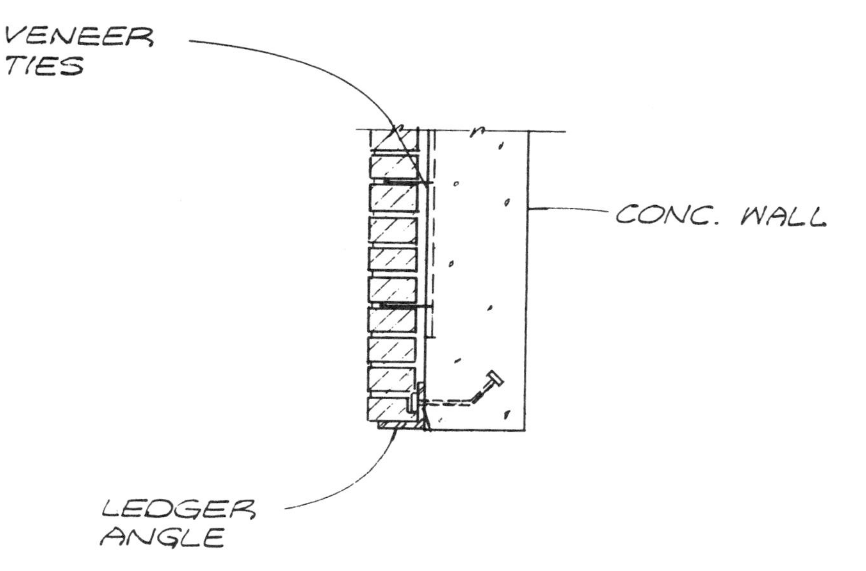

Detail 54. A section of a concrete wall lintel supporting a brick veneer. The veneer is supported vertically by a continuous angle bolted to the concrete as shown. The size and spacing of the angle bolts are determined by calculation and by the local building code.

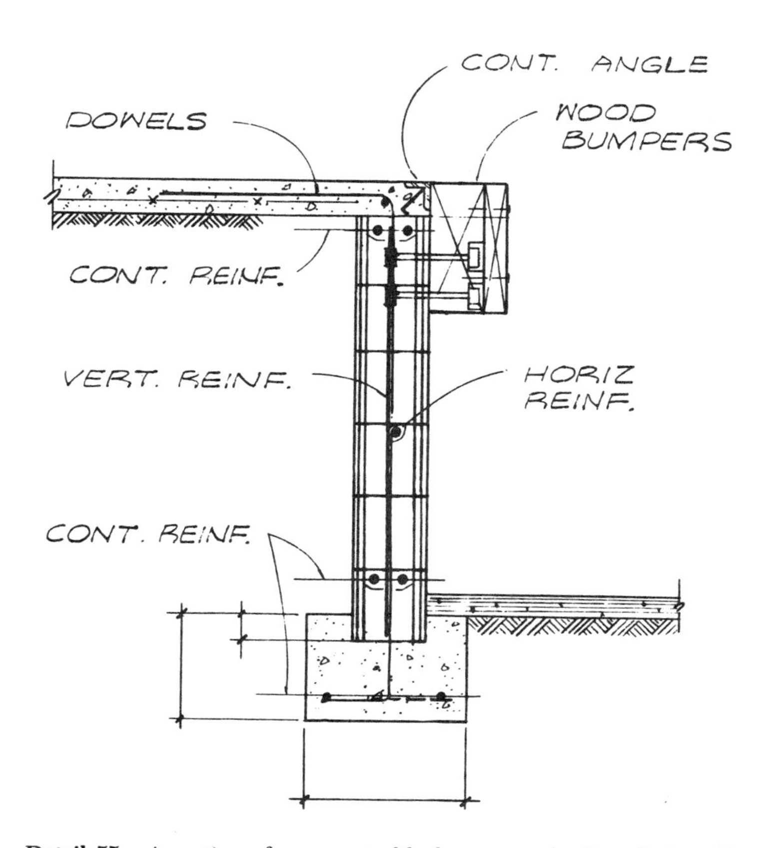

Detail 55. A section of a concrete block masonry loading dock wall. The wall is connected to the concrete slab on grade with reinforcing dowels. The vertical and horizontal reinforcing steel in the wall is determined by calculation. The depth and width of the wall footing depend on the soil conditions. The wood bumpers are bolted to the face of the masonry wall. The continuous angle at the top of the concrete slab prevents the concrete from spalling.

Notes ▪ Drawings ▪ Ideas

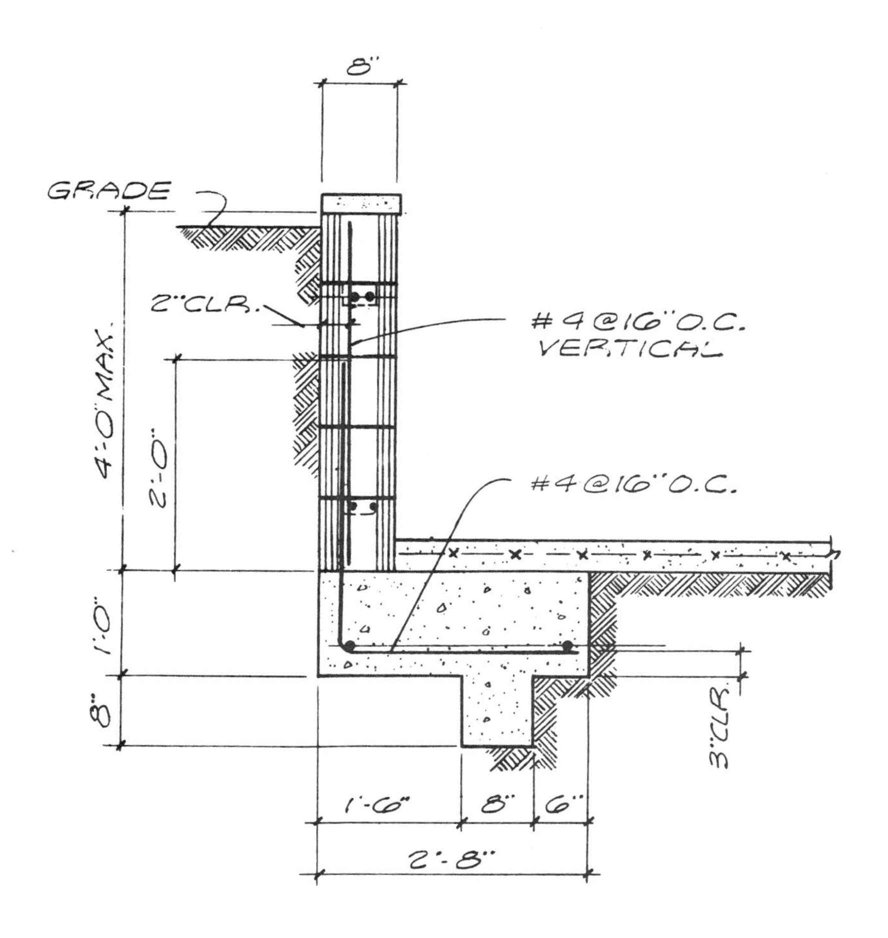

Detail 56(a). Stem height 4′0″.

Detail 56. Sections of concrete block masonry retaining walls with soil at the exterior face of the stem. The walls retain a flat grade with no surcharge. The wall dimensions and the reinforcing steel size and spacing are determined by calculation. All concrete block masonry cells are filled with grout. Two #4 horizontal continuous reinforcing bars are placed at the top and the bottom of the wall stem. The minimum horizontal reinforcing in the stem wall and the footing is #4 at 24″ o.c. The masonry head joint is omitted at 32″ o.c. at the first course as a weep hole. The cement cap at the top of the wall is optional. The walls are designed for 30 lb per cu ft equivalent fluid pressure and a maximum soil bearing pressure of 1500 lb per sq ft. The resultant of the forces passes through the middle ⅓ of the wall footing. The walls are designed to resist sliding and overturning. The overturning safety factor is 1.5. The concrete design stress is $f'c = 2000$ psi. The masonry design stress is $f'm = 600$ psi.

Notes ▪ Drawings ▪ Ideas

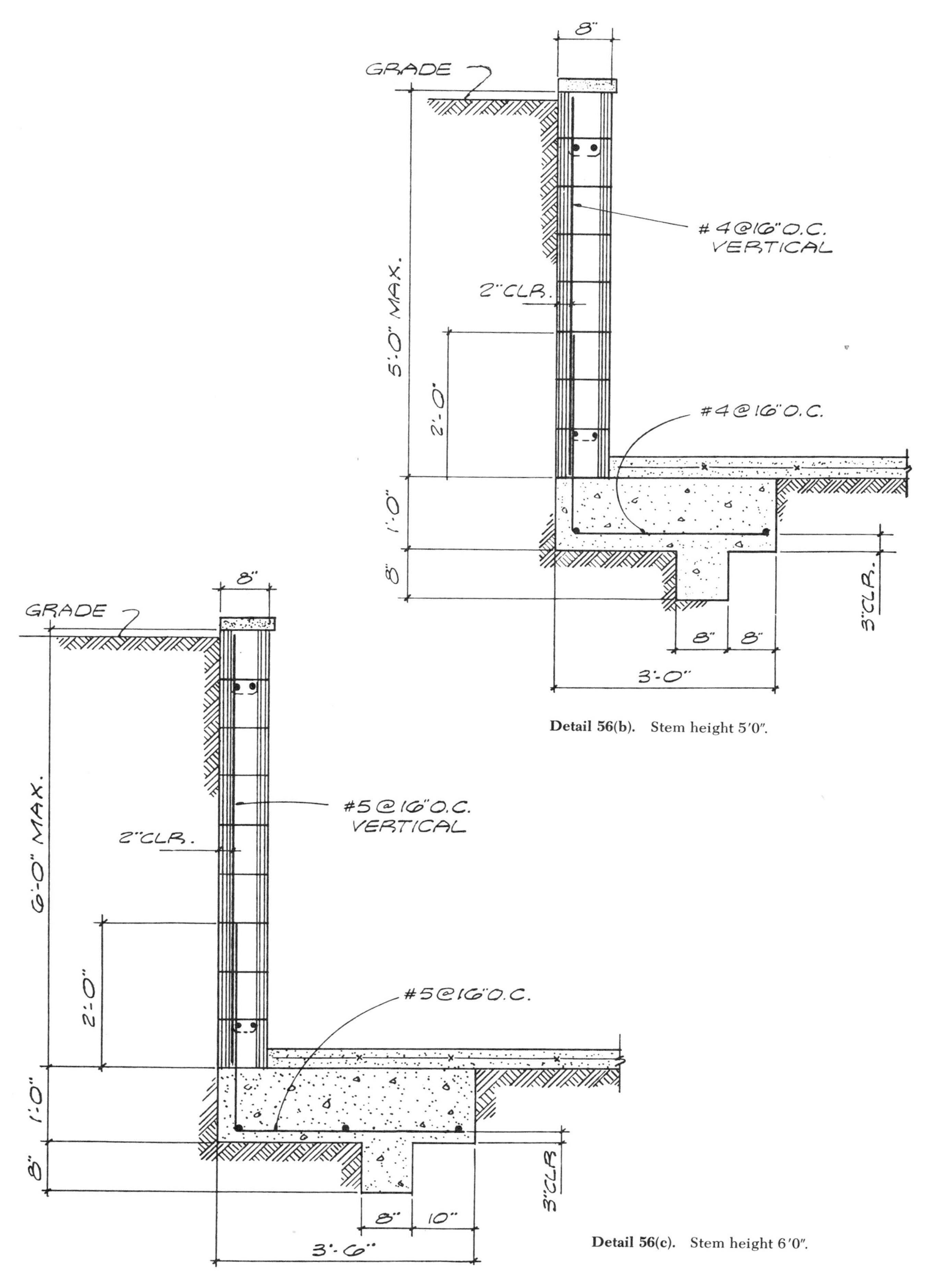

Detail 56(b). Stem height 5′0″.

Detail 56(c). Stem height 6′0″.

Notes ▪ Drawings ▪ Ideas

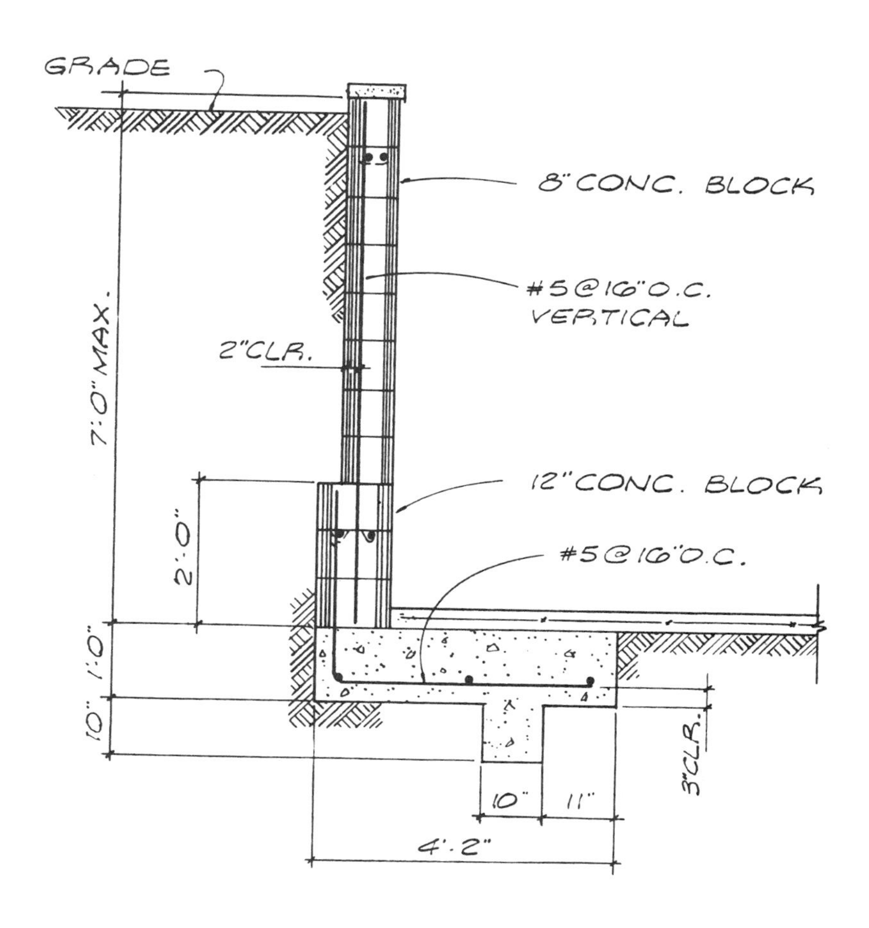

Detail 56(d). Stem height 7′0″.

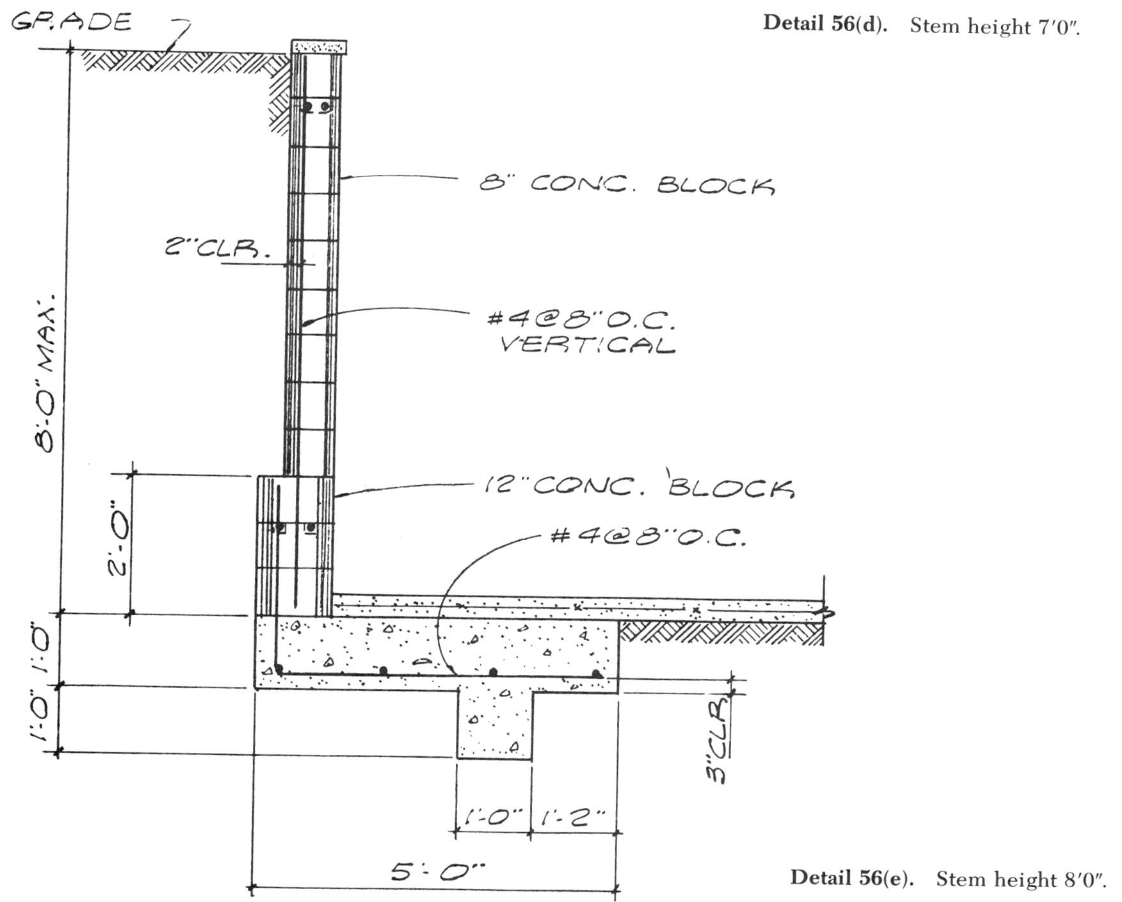

Detail 56(e). Stem height 8′0″.

Notes ▪ Drawings ▪ Ideas

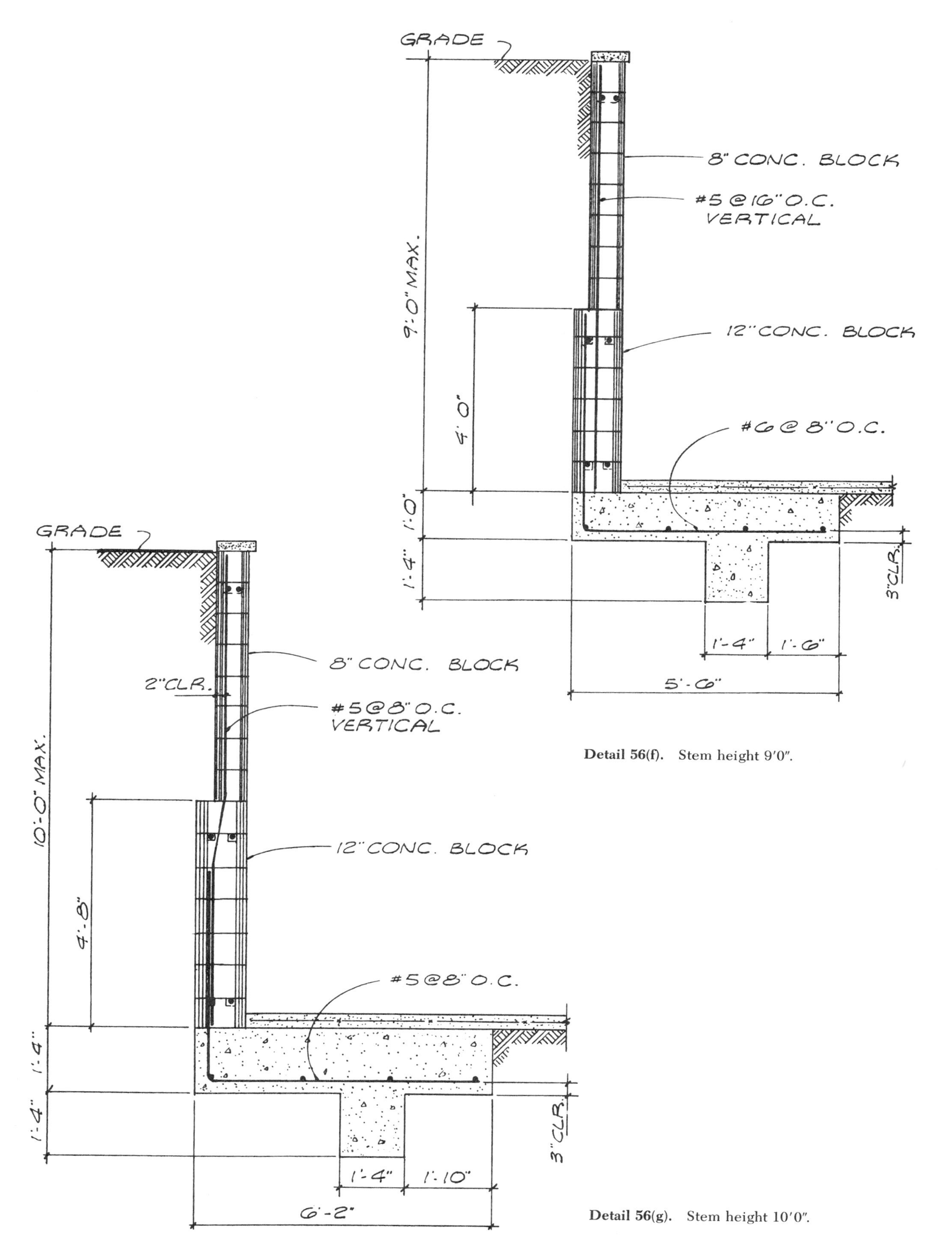

Detail 56(f). Stem height 9′0″.

Detail 56(g). Stem height 10′0″.

Notes ▪ Drawings ▪ Ideas

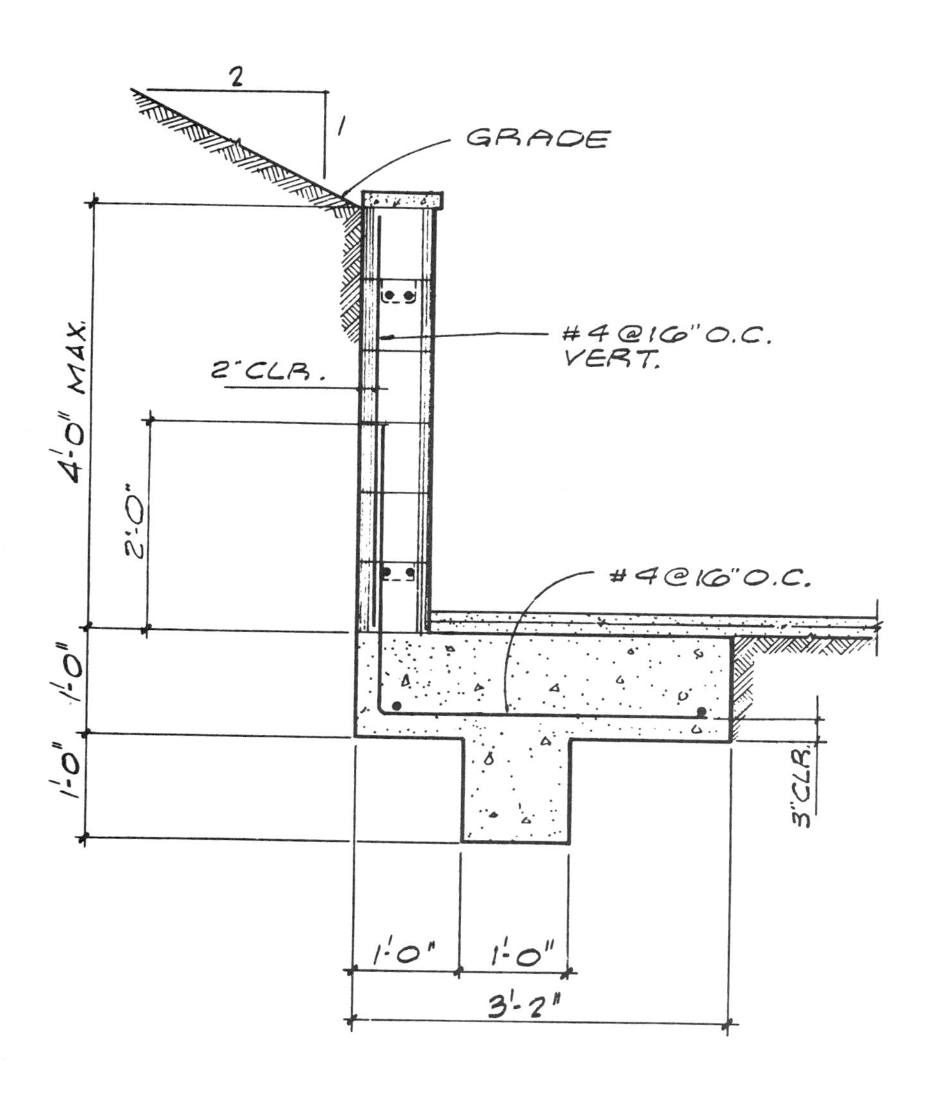

Detail 57(a). Stem height 4′0″.

Detail 57. Sections of concrete block masonry retaining walls with soil at the exterior face of the stem. The wall retains a sloping grade of two horizontal to one vertical and no surcharge. The wall dimensions and reinforcing steel size and spacing are determined by calculation. All concrete block masonry cells are filled with grout. Two #4 horizontal continuous reinforcing bars are placed at the top and the bottom of the wall stem. The minimum horizontal reinforcing in the stem wall and the footing is #4 at 24″ o.c. The masonry head joint is omitted at 32″ o.c. at the first course as a weep hole. The cement cap at the top of the wall is optional. The walls are designed for 43 lb per cu ft equivalent fluid pressure and a maximum soil bearing pressure of 1500 lb per sq ft. The resultant of the forces passes through the middle ⅓ of the wall footing. The walls are designed to resist sliding and overturning. The overturning safety factor is 1.5. The concrete design stress is $f'c = 2000$ psi. The masonry design stress is $f'm = 600$ psi.

Notes ▪ Drawings ▪ Ideas

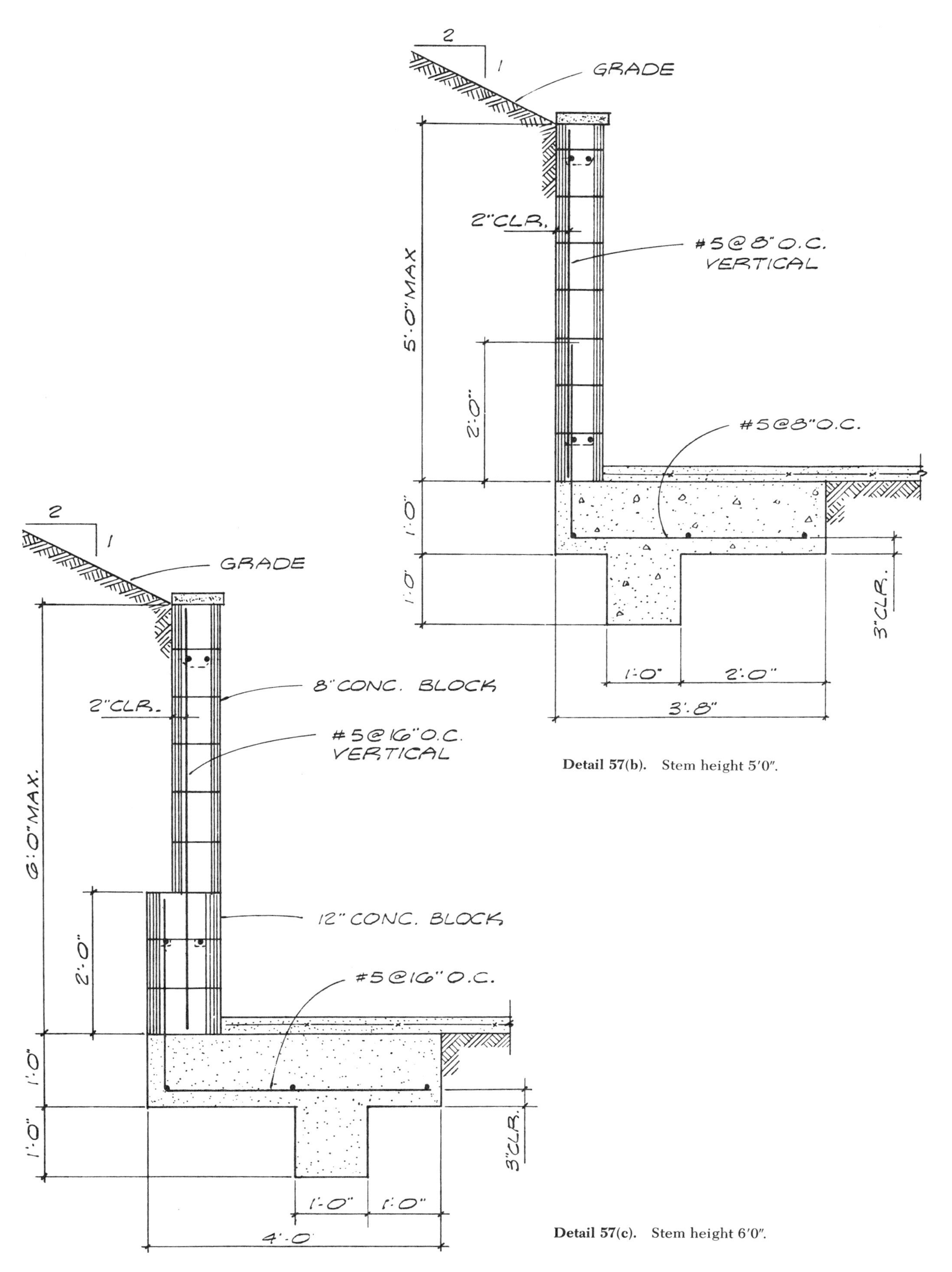

Detail 57(b). Stem height 5′0″.

Detail 57(c). Stem height 6′0″.

Notes ▪ Drawings ▪ Ideas

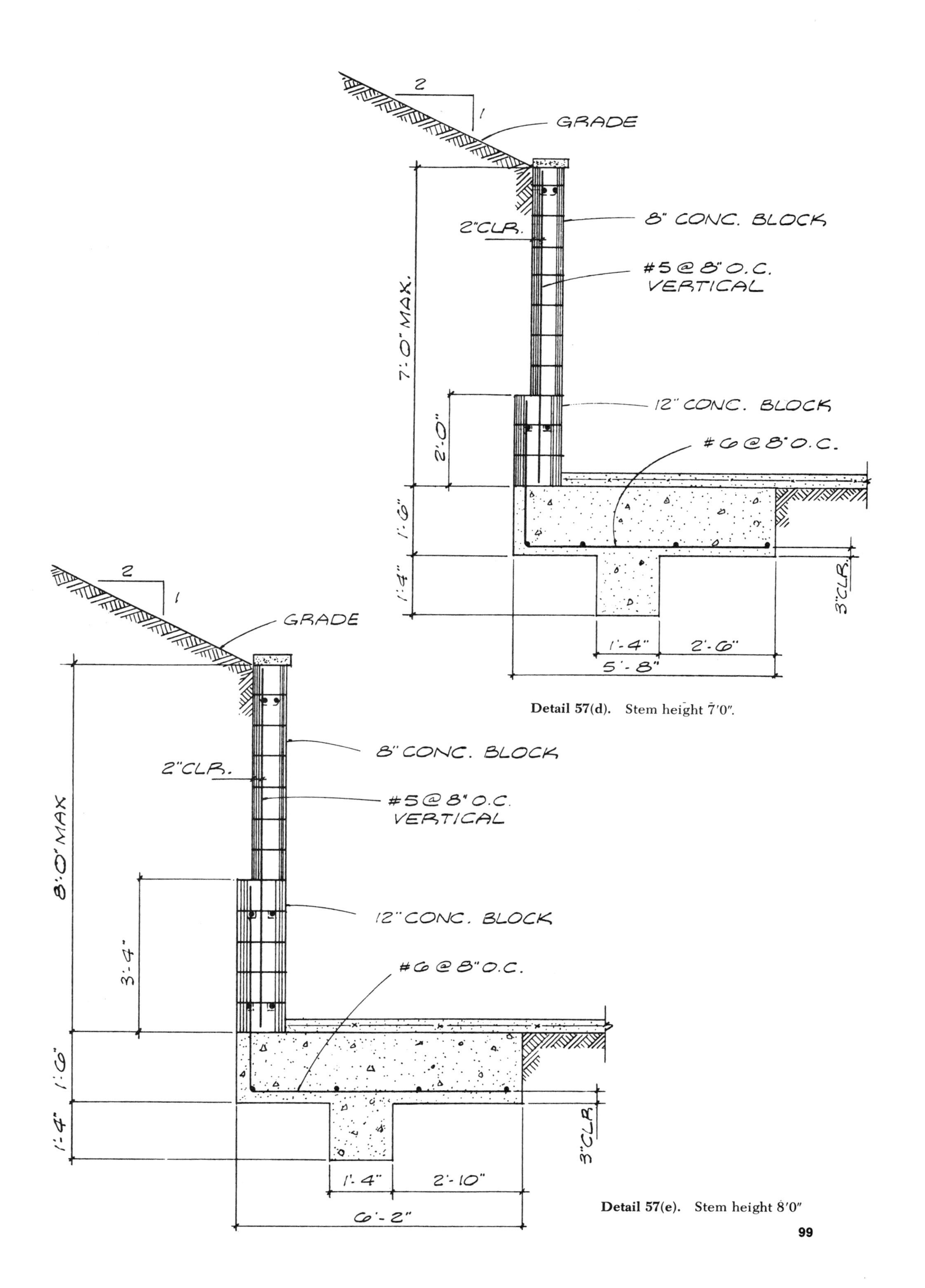

Detail 57(d). Stem height 7'0".

Detail 57(e). Stem height 8'0"

Notes ▪ Drawings ▪ Ideas

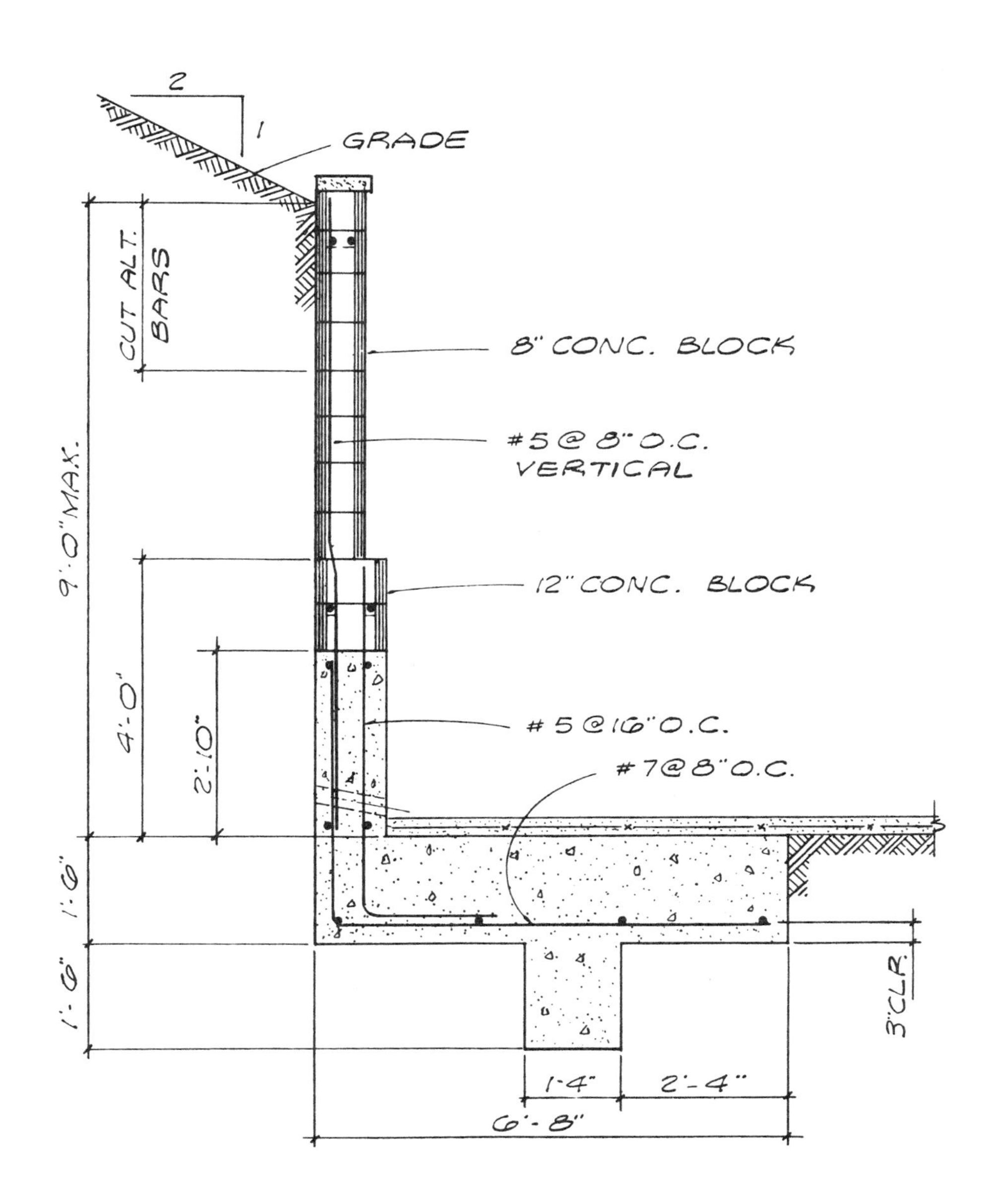

Detail 57(f). Stem height 9′0″.

Notes ▪ Drawings ▪ Ideas

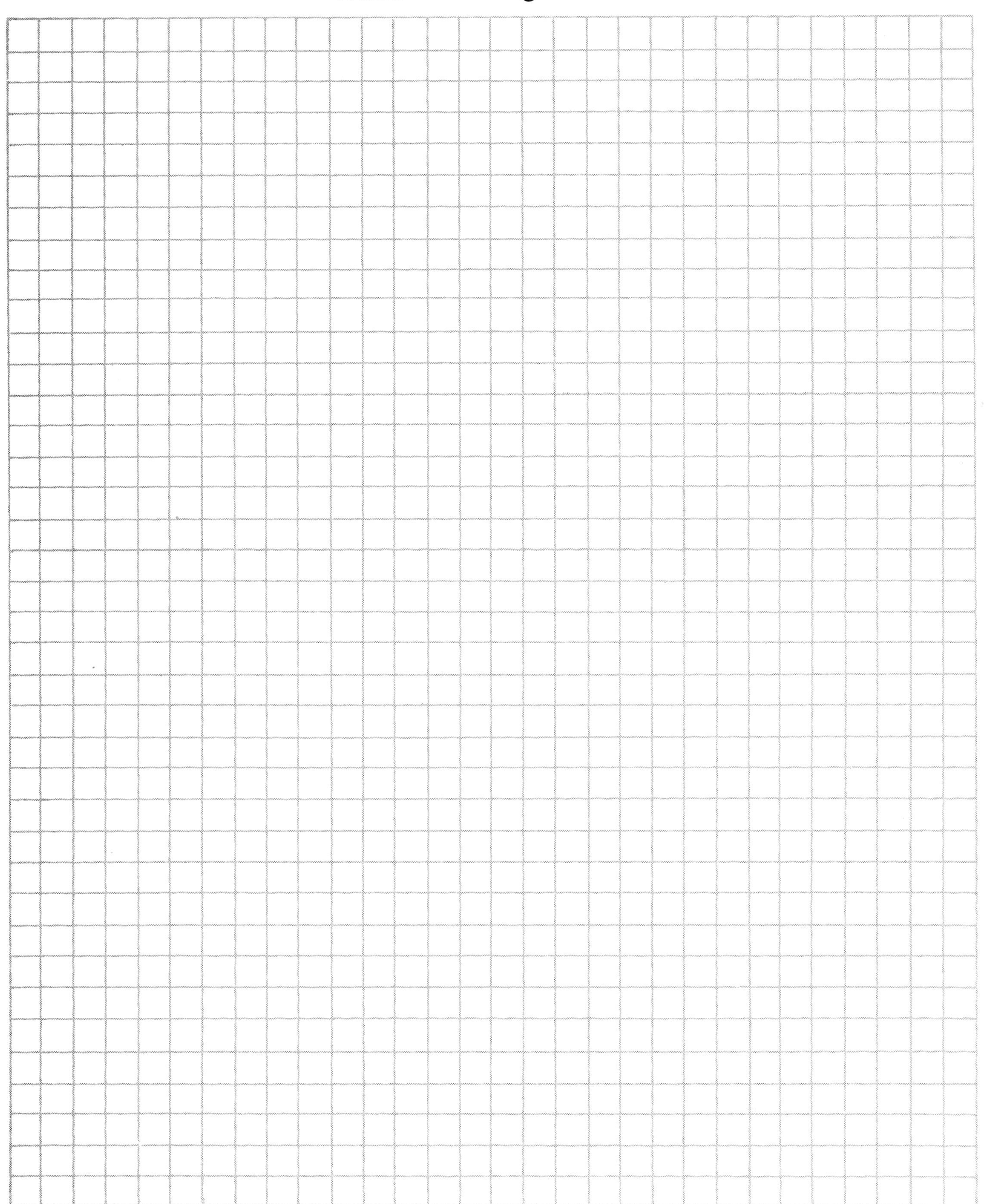

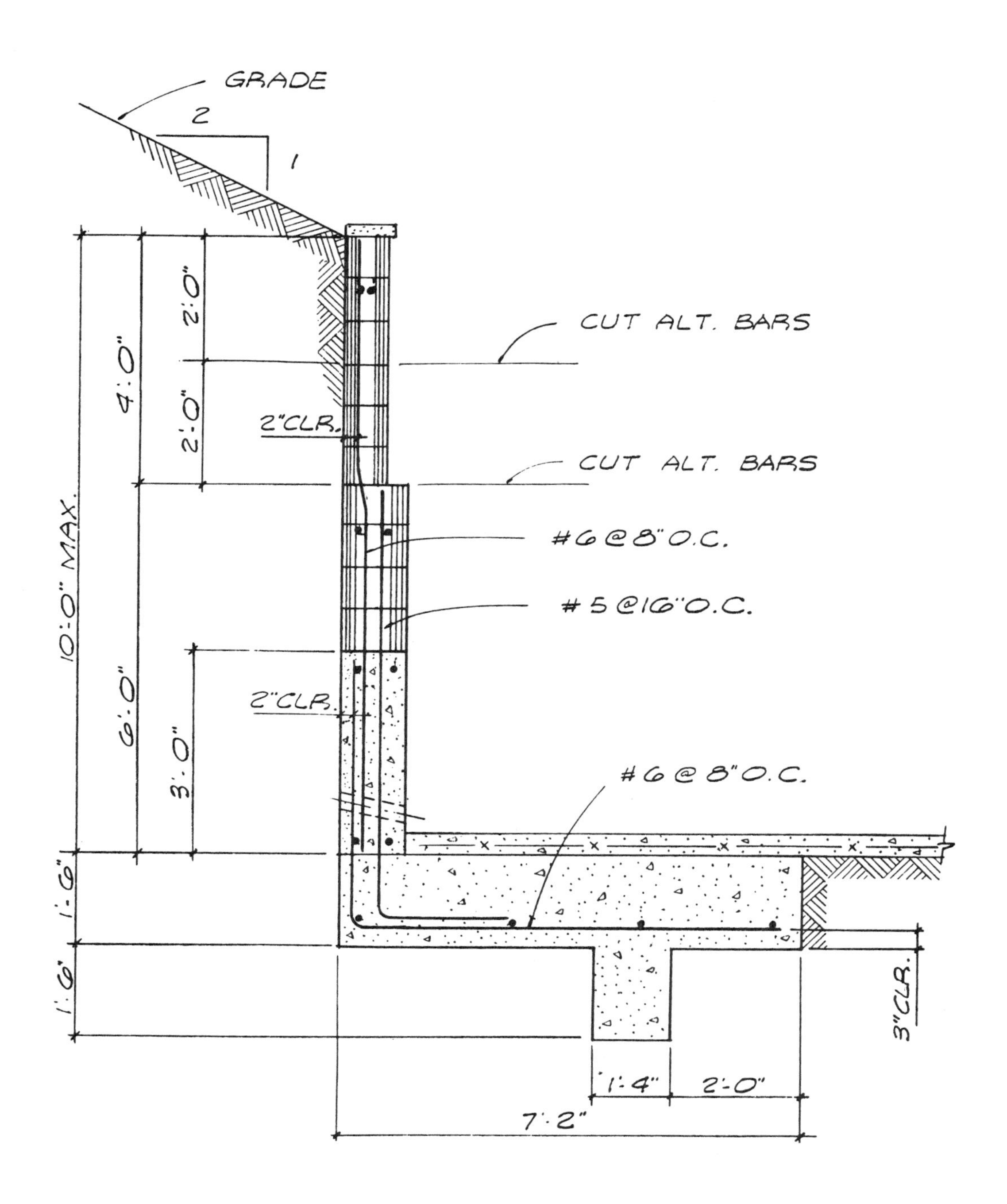

Detail 57(g). Stem height 10′0″.

Notes ▪ Drawings ▪ Ideas

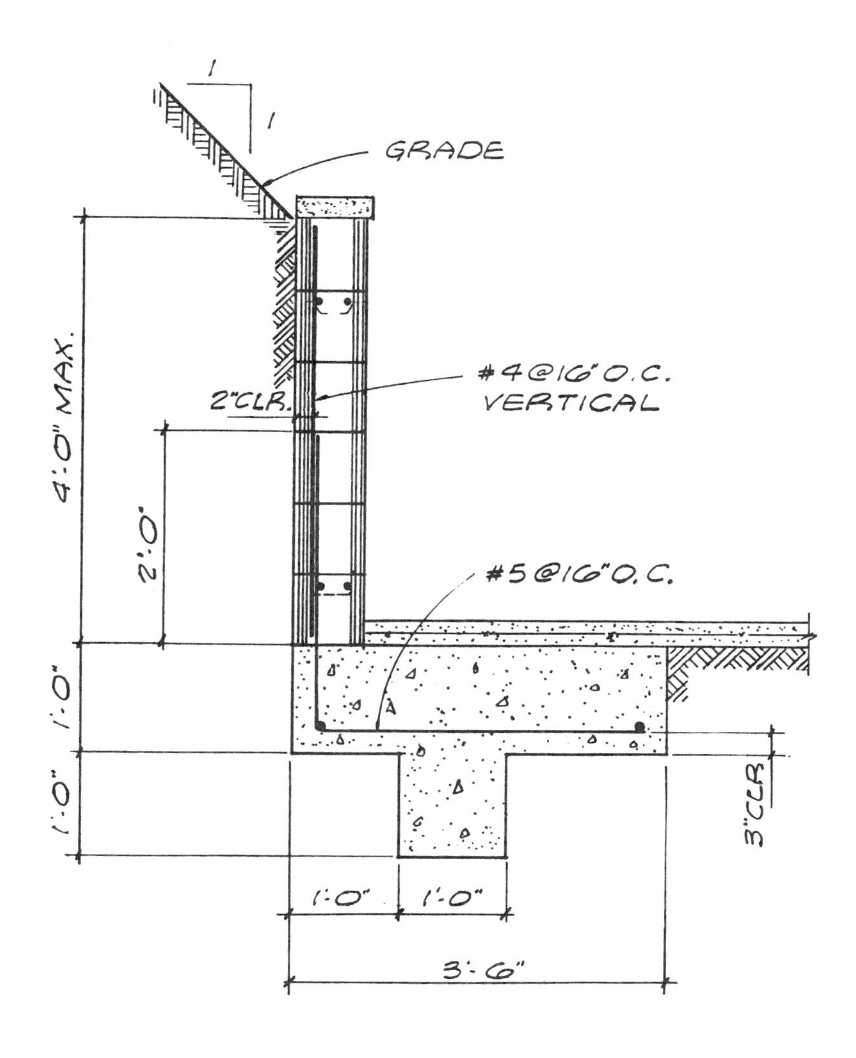

Detail 58(a). Stem height 4′0″.

Detail 58. Sections of concrete block masonry retaining walls with soil at the exterior face of the stem. The wall retains a sloping grade with one horizontal to one vertical and no surcharge. The wall dimensions and reinforcing steel size and spacing are determined by calculation. All concrete block masonry cells are filled with grout. Two #4 horizontal continuous reinforcing bars are placed at the top and the bottom of the wall stem. The minimum horizontal reinforcing in the stem wall and the footing is #4 at 24″ o.c. The masonry head joint is omitted at 32″ o.c. at the first course as a weep hole. The cement cap at the top of the wall is optional. The walls are designed for 80 lb per cu ft equivalent fluid pressure and a maximum soil bearing pressure of 1500 lb per sq ft. The resultant of the forces passes through the middle ⅓ of the wall footing. The walls are designed to resist sliding and overturning. The overturning safety factor is 1.5. The concrete design stress is $f'c = 2000$ psi. The masonry design stress is $f'm = 600$ psi.

Notes ▪ Drawings ▪ Ideas

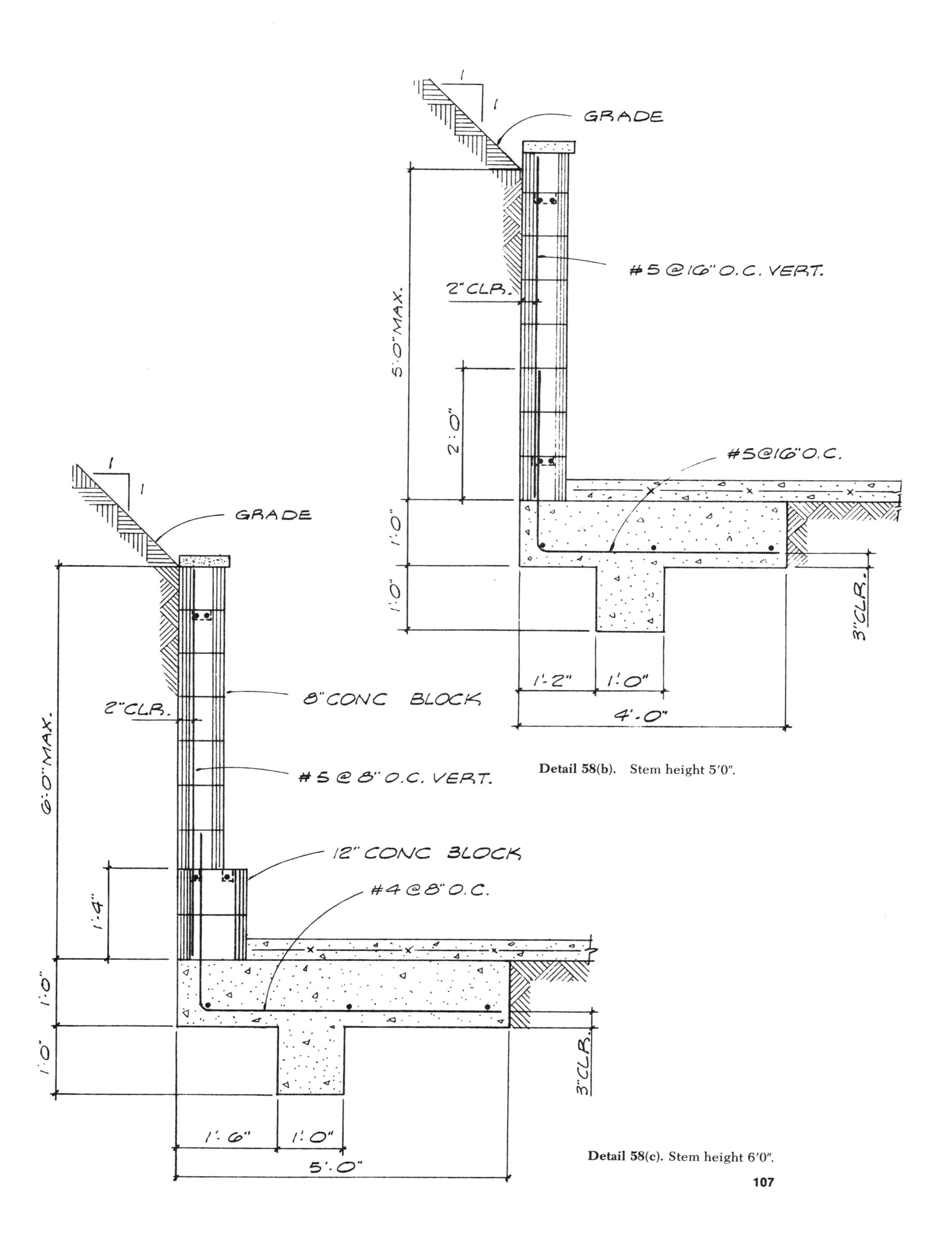

Detail 58(b). Stem height 5′0″.

Detail 58(c). Stem height 6′0″.

Notes ▪ Drawings ▪ Ideas

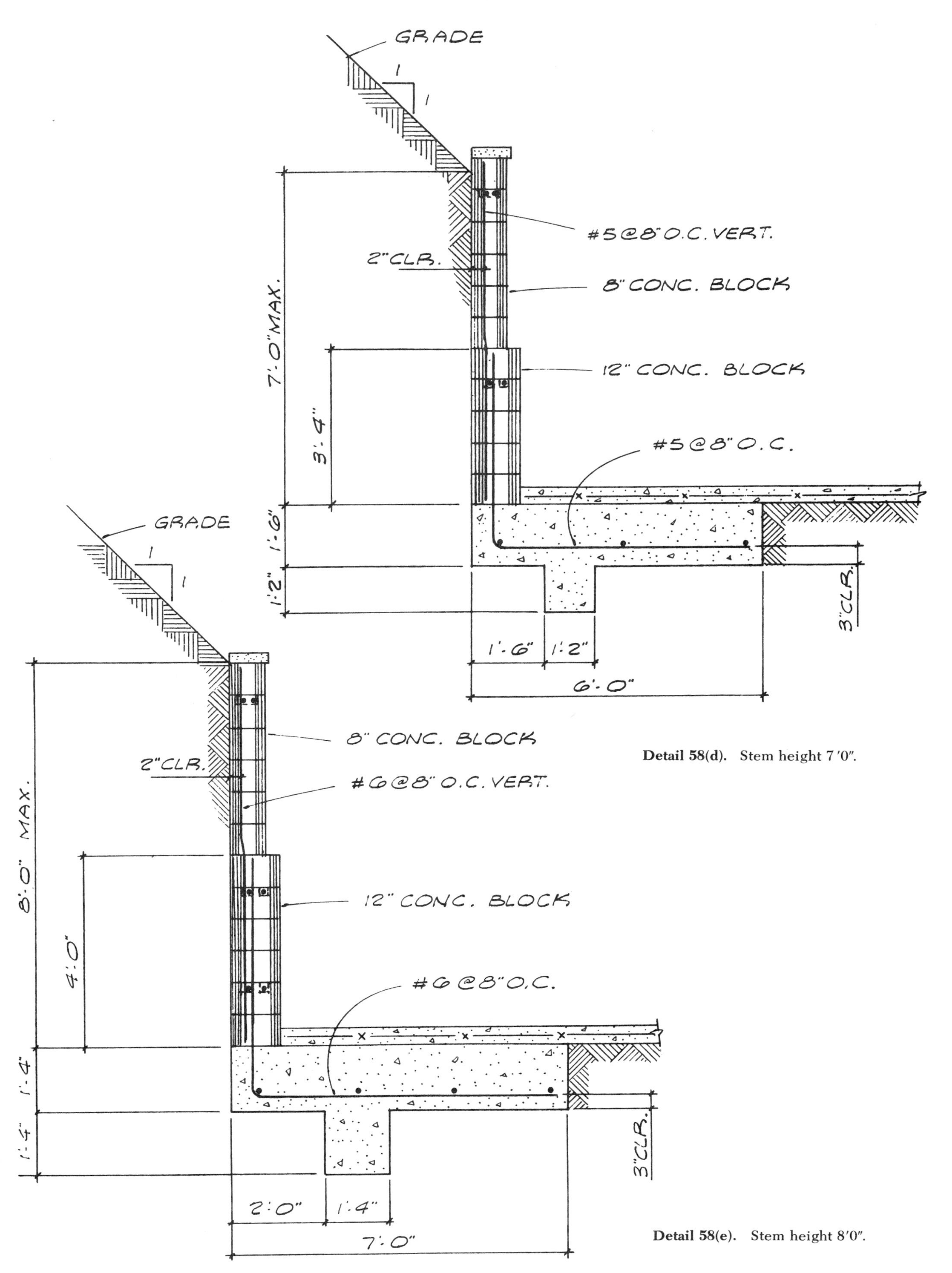

Detail 58(d). Stem height 7'0".

Detail 58(e). Stem height 8'0".

Notes ▪ Drawings ▪ Ideas

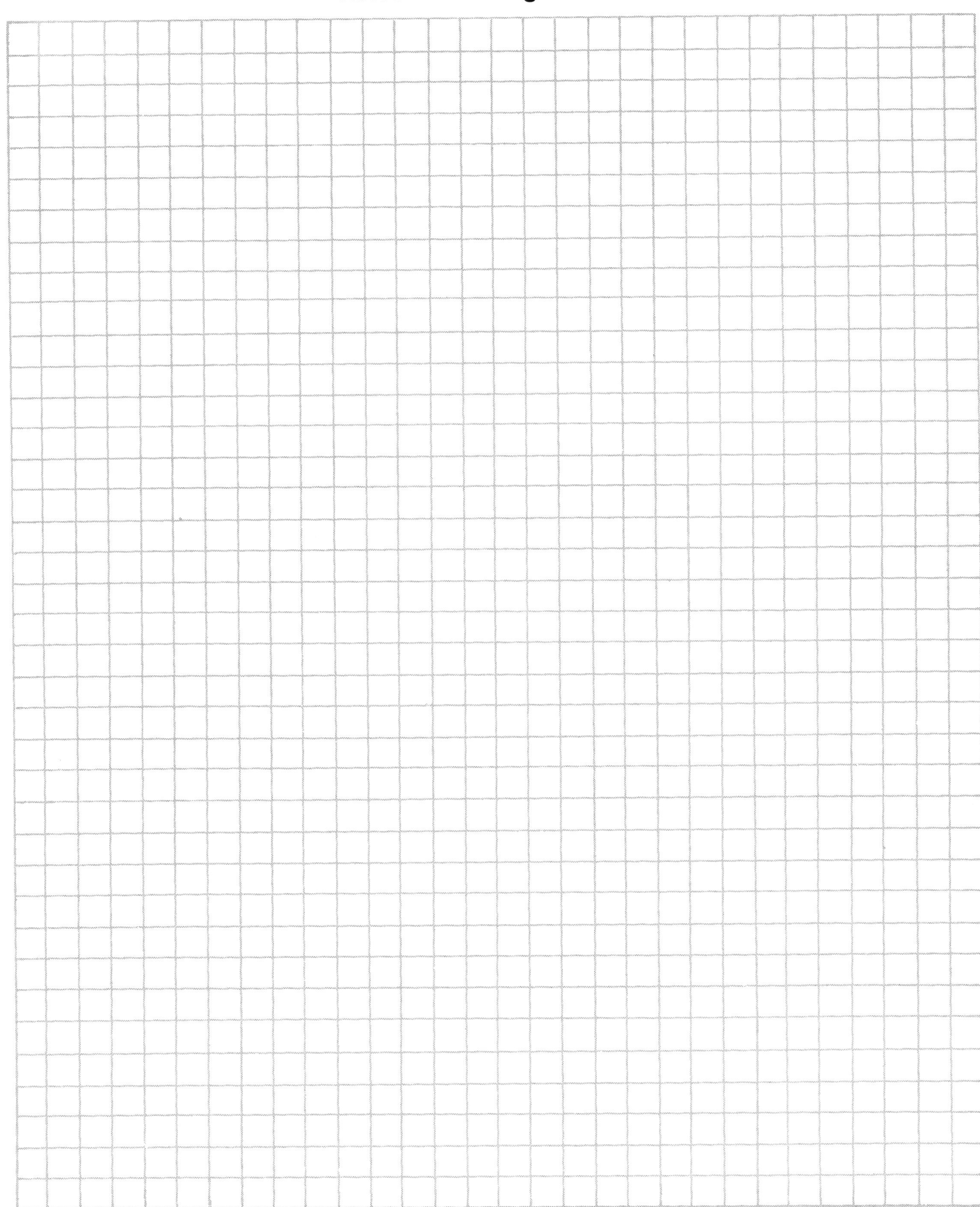

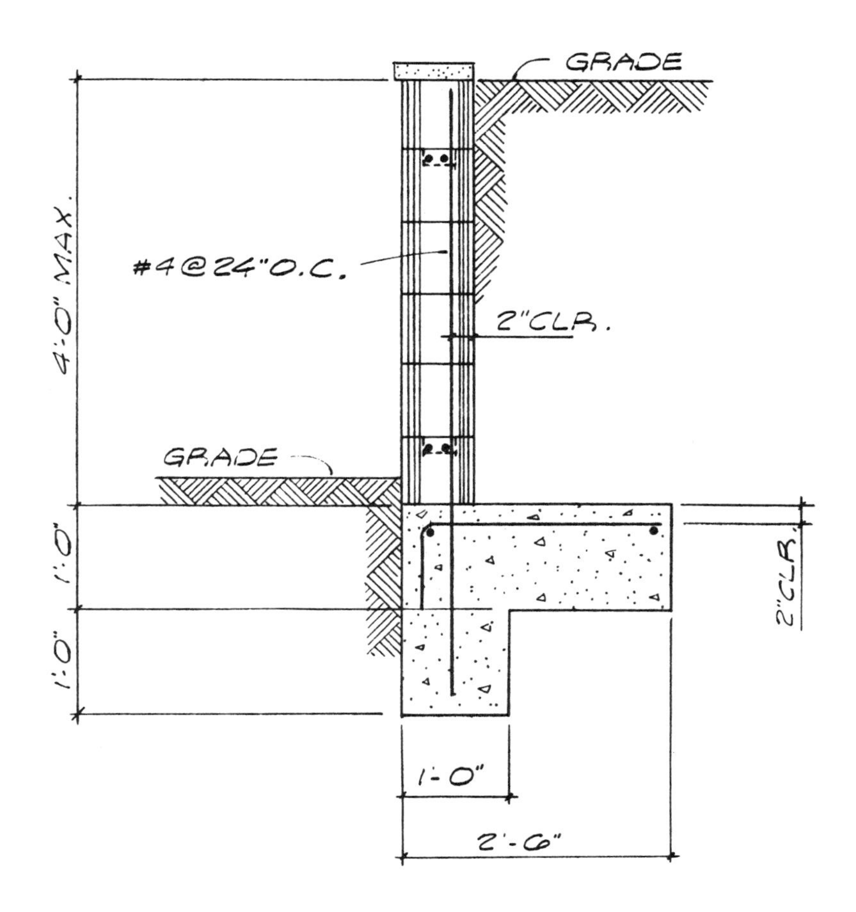

Detail 59(a). Stem height 4′0″.

Detail 59. Sections of concrete block masonry retaining walls with soil at the interior face of the stem. The walls retain a flat grade with no surcharge. The wall dimensions and reinforcing steel size and spacing are determined by calculation. All concrete block masonry cells are filled with grout. Two #4 horizontal continuous reinforcing bars are placed at the top and the bottom of the wall stem. The minimum horizontal reinforcing in the stem wall and the footing is #4 at 24″ o.c. The masonry head joint is omitted at 32″ o.c. at the first course as a weep hole. The cement cap at the top of the wall is optional. The walls are designed for 30 lb per cu ft equivalent fluid pressure and a maximum soil bearing pressure of 1500 lb per sq ft. The resultant of the forces passes through the middle ⅓ of the wall footing. The walls are designed to resist sliding and overturning. The overturning safety factor is 1.5. The concrete design stress is $f'c = 2000$ psi. The masonry design stress is $f'm = 600$ psi.

Notes ▪ Drawings ▪ Ideas

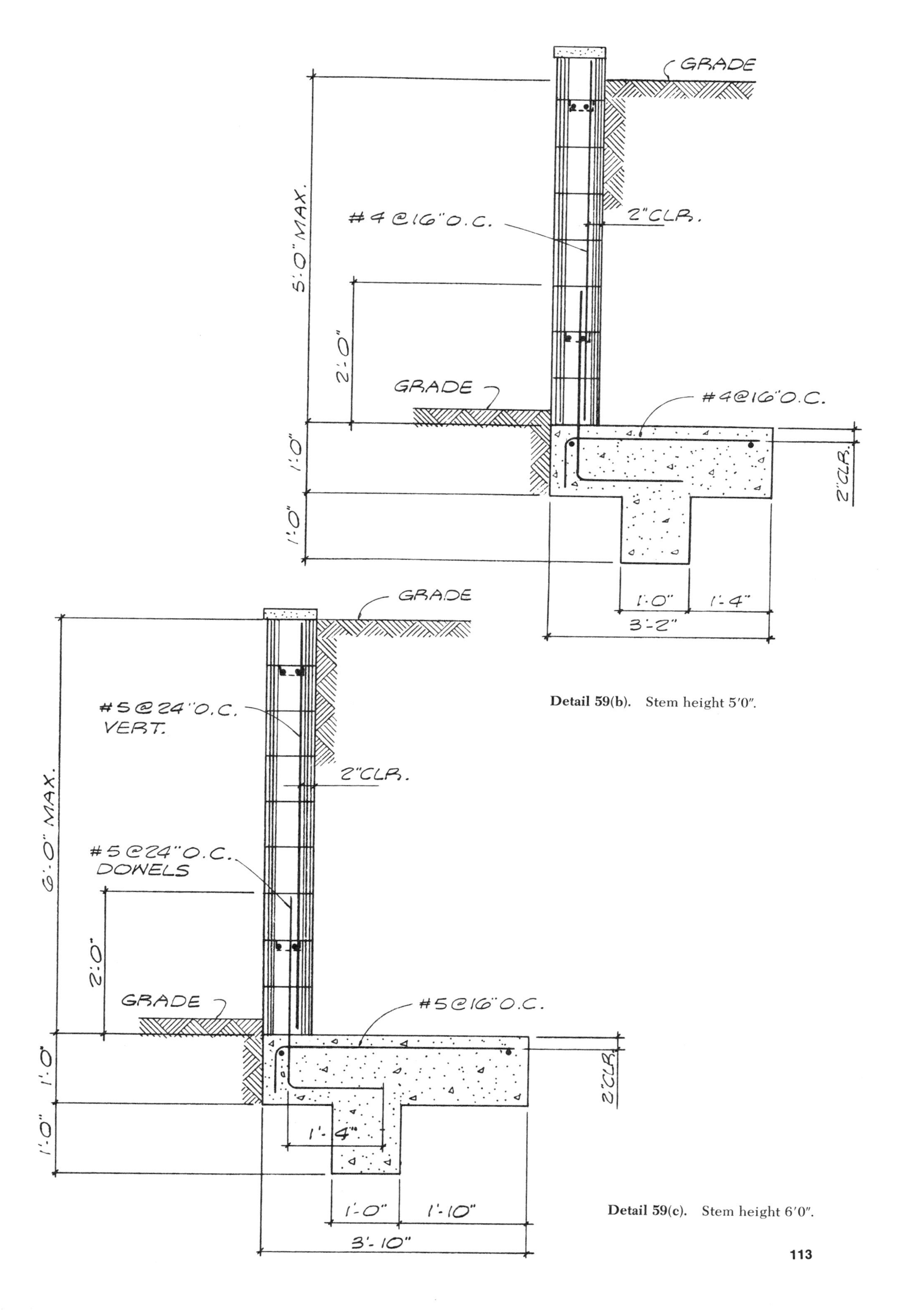

Detail 59(b). Stem height 5′0″.

Detail 59(c). Stem height 6′0″.

Notes ▪ Drawings ▪ Ideas

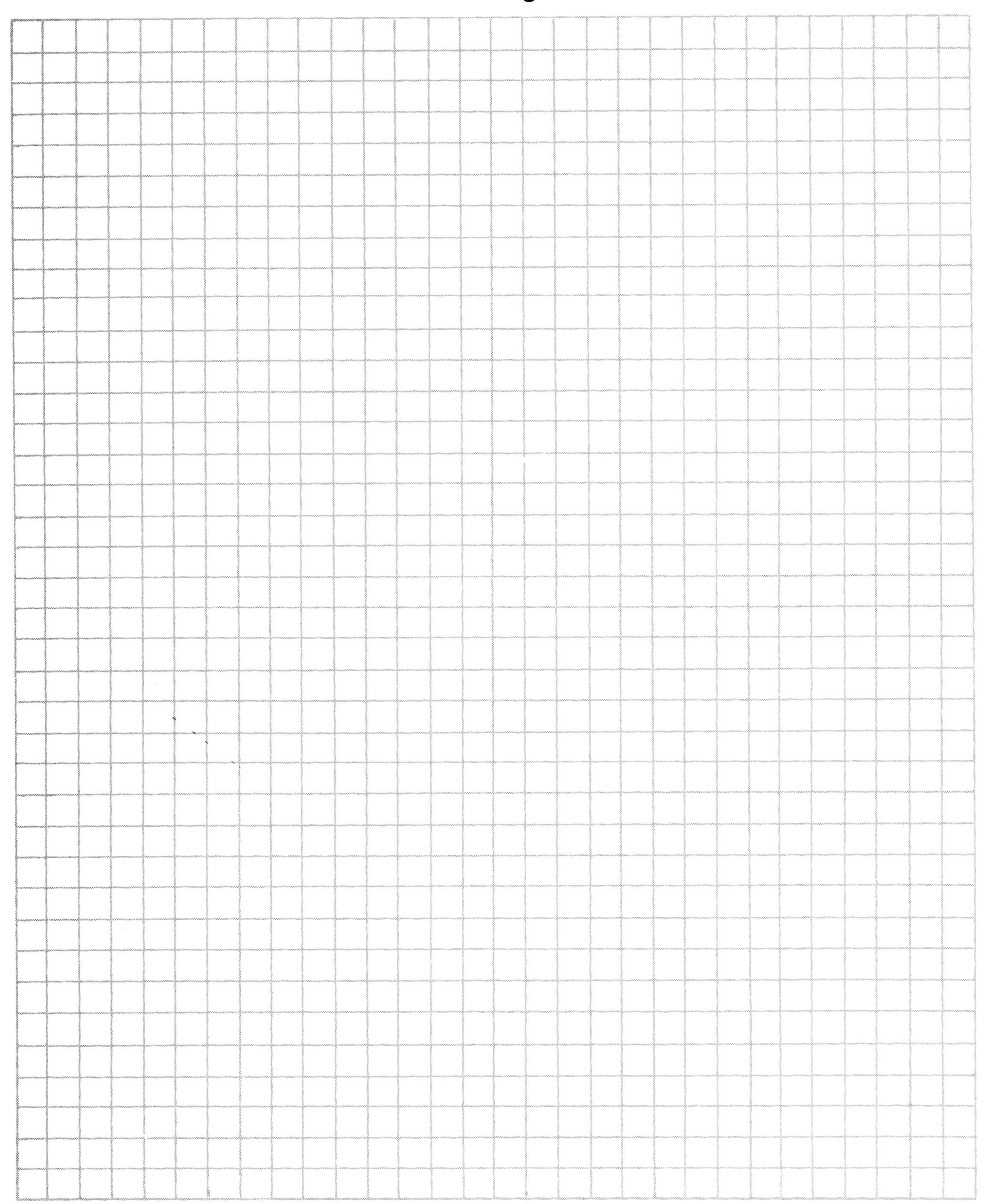

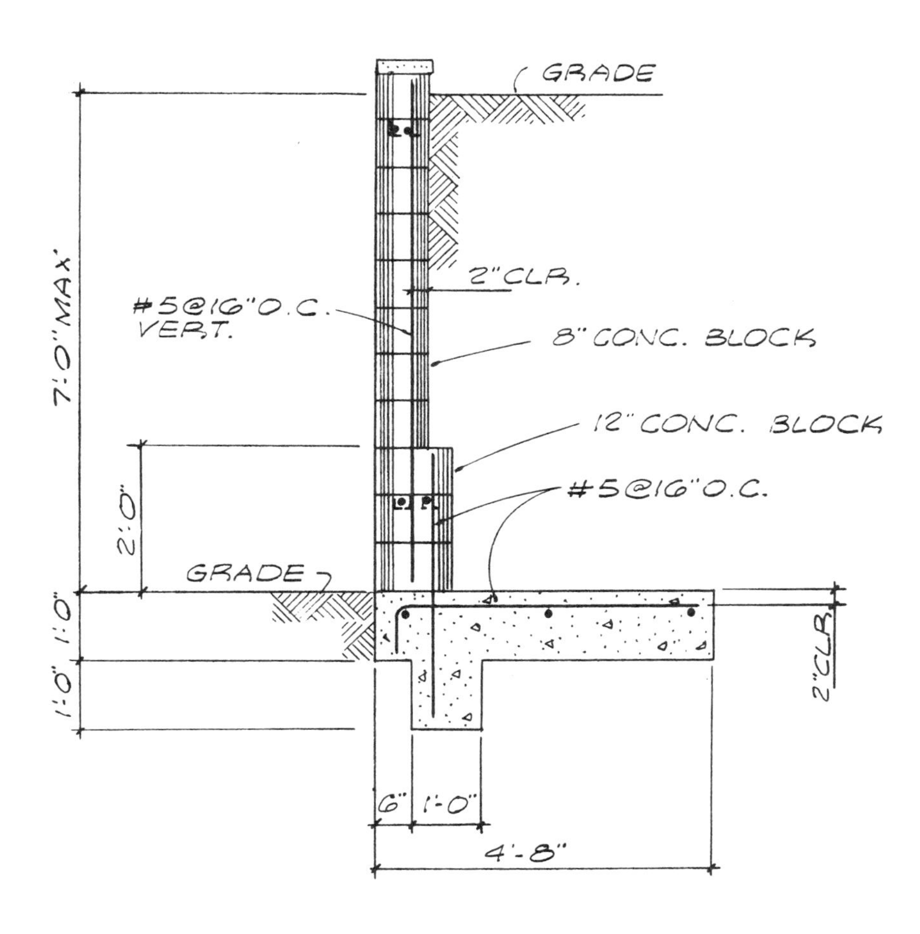

Detail 59(d). Stem height 7′0″.

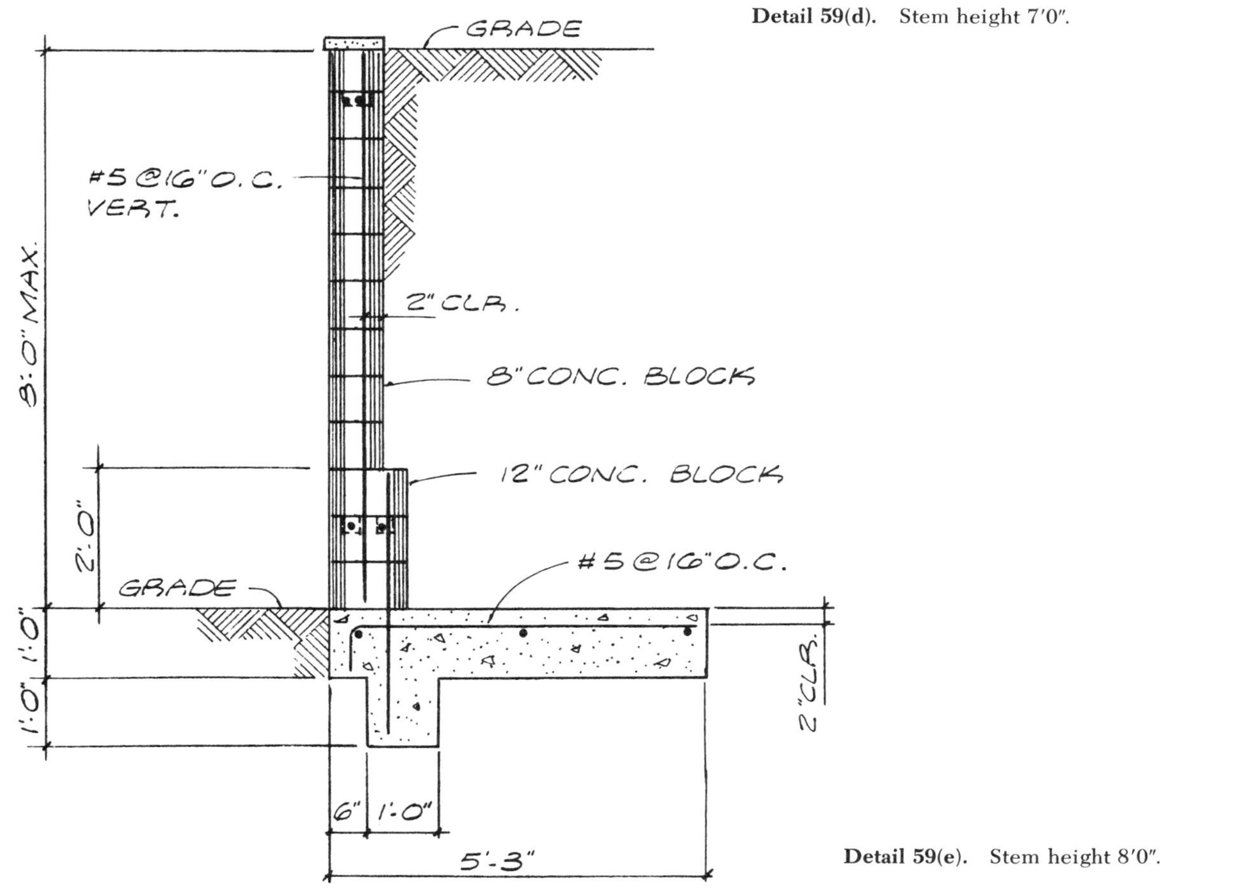

Detail 59(e). Stem height 8′0″.

Notes ▪ Drawings ▪ Ideas

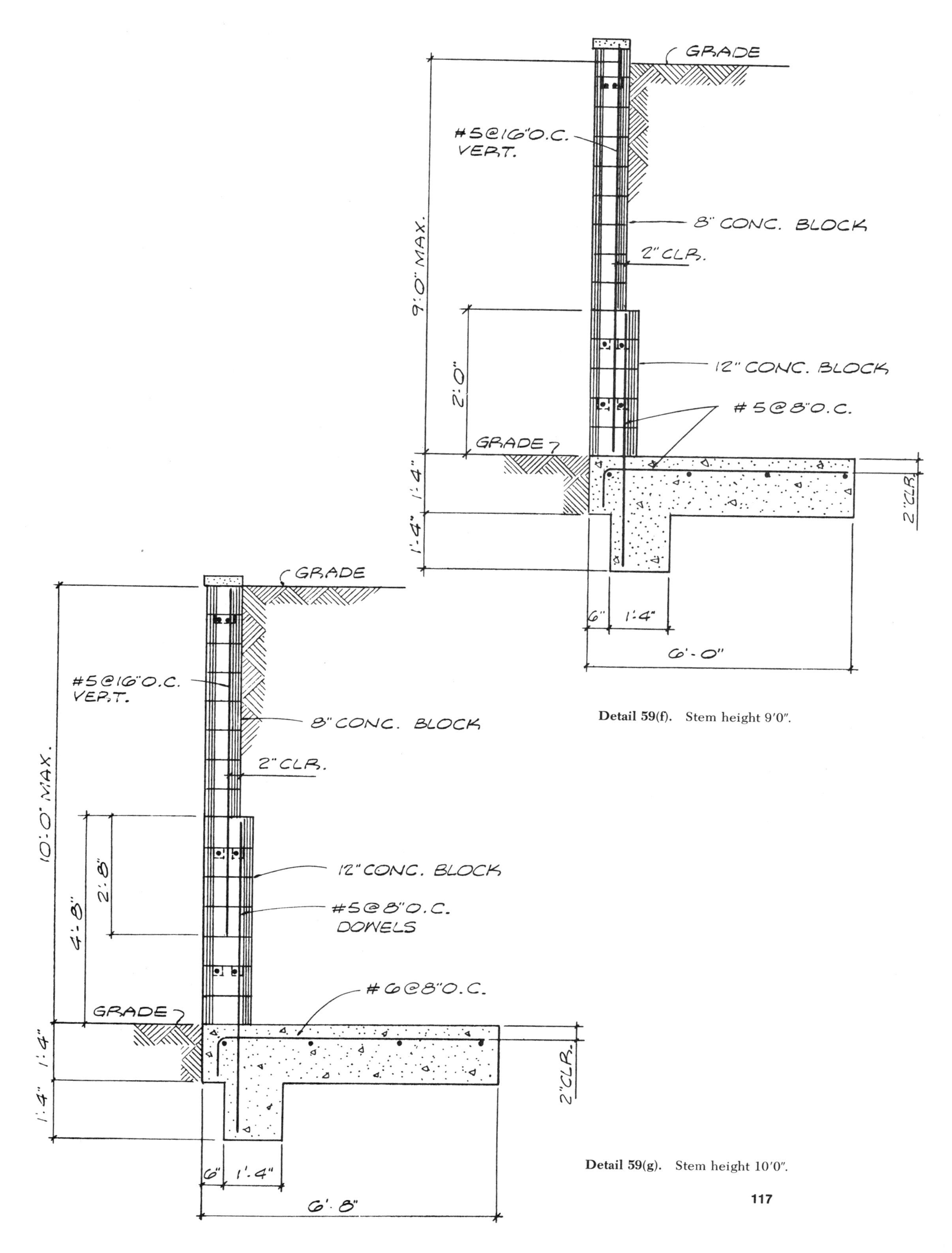

Detail 59(f). Stem height 9′0″.

Detail 59(g). Stem height 10′0″.

Notes ▪ Drawings ▪ Ideas

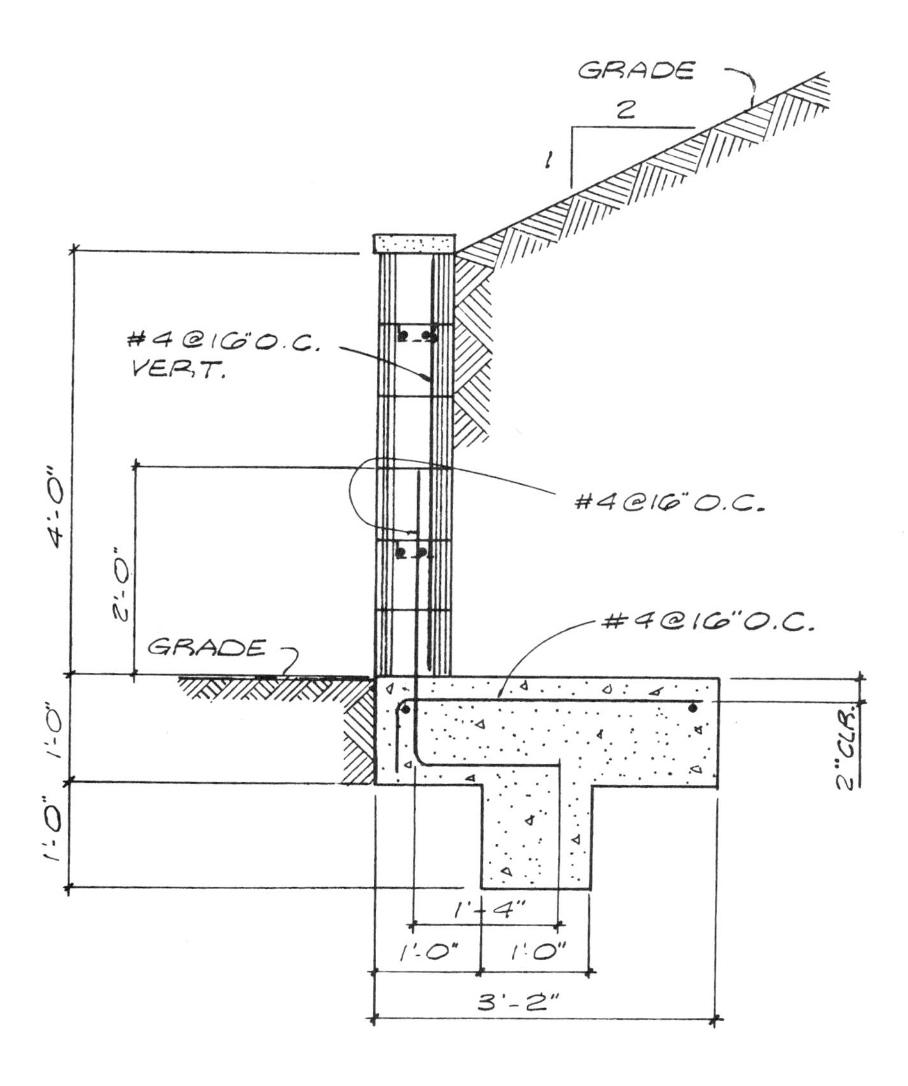

Detail 60(a). Stem height 4′0″.

Detail 60. Sections of concrete block masonry retaining walls with soil at the interior face of the stem. The wall retains a sloping grade of two horizontal to one vertical and no surcharge. The wall dimensions and reinforcing steel size and spacing are determined by calculation. All concrete block masonry cells are filled with grout. Two #4 horizontal continuous reinforcing bars are placed at the top and the bottom of the wall stem. The minimum horizontal reinforcing in the stem wall and the footing is #4 at 24″ o.c. The masonry head joint is omitted at 32″ o.c. at the first course as a weep hole. The cement cap at the top of the wall is optional. The walls are designed for 43 lb per cu ft equivalent fluid pressure and a maximum soil bearing pressure of 1500 lb per sq. ft. The resultant of the forces passes through the middle ⅓ of the wall footing. The walls are designed to resist sliding and overturning. The overturning safety factor is 1.5. The concrete design stress is $f'c = 2000$ psi. The masonry design stress is $f'm = 600$ psi.

Notes ▪ Drawings ▪ Ideas

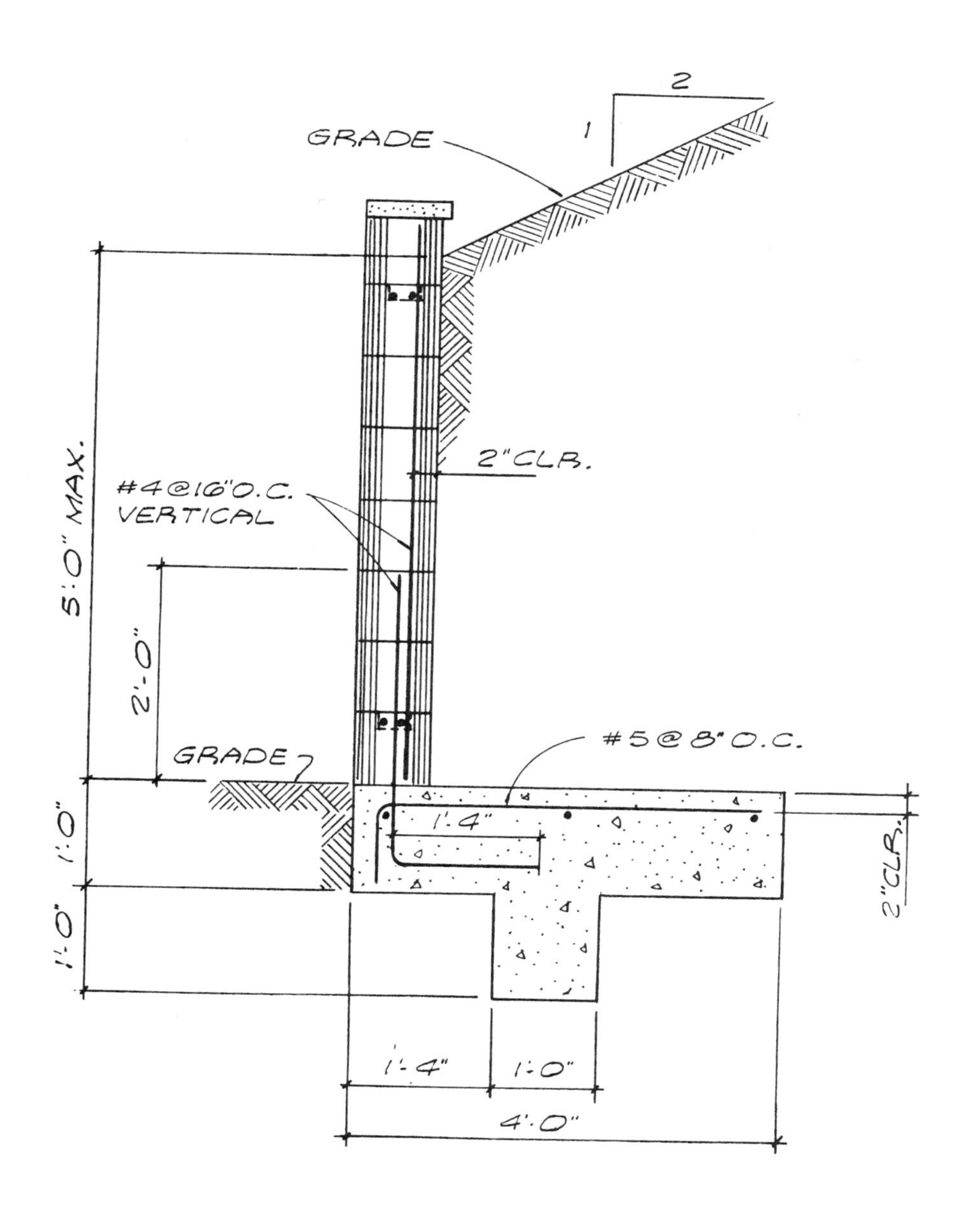

Detail 60(b). Stem height 5′0″.

Notes ▪ Drawings ▪ Ideas

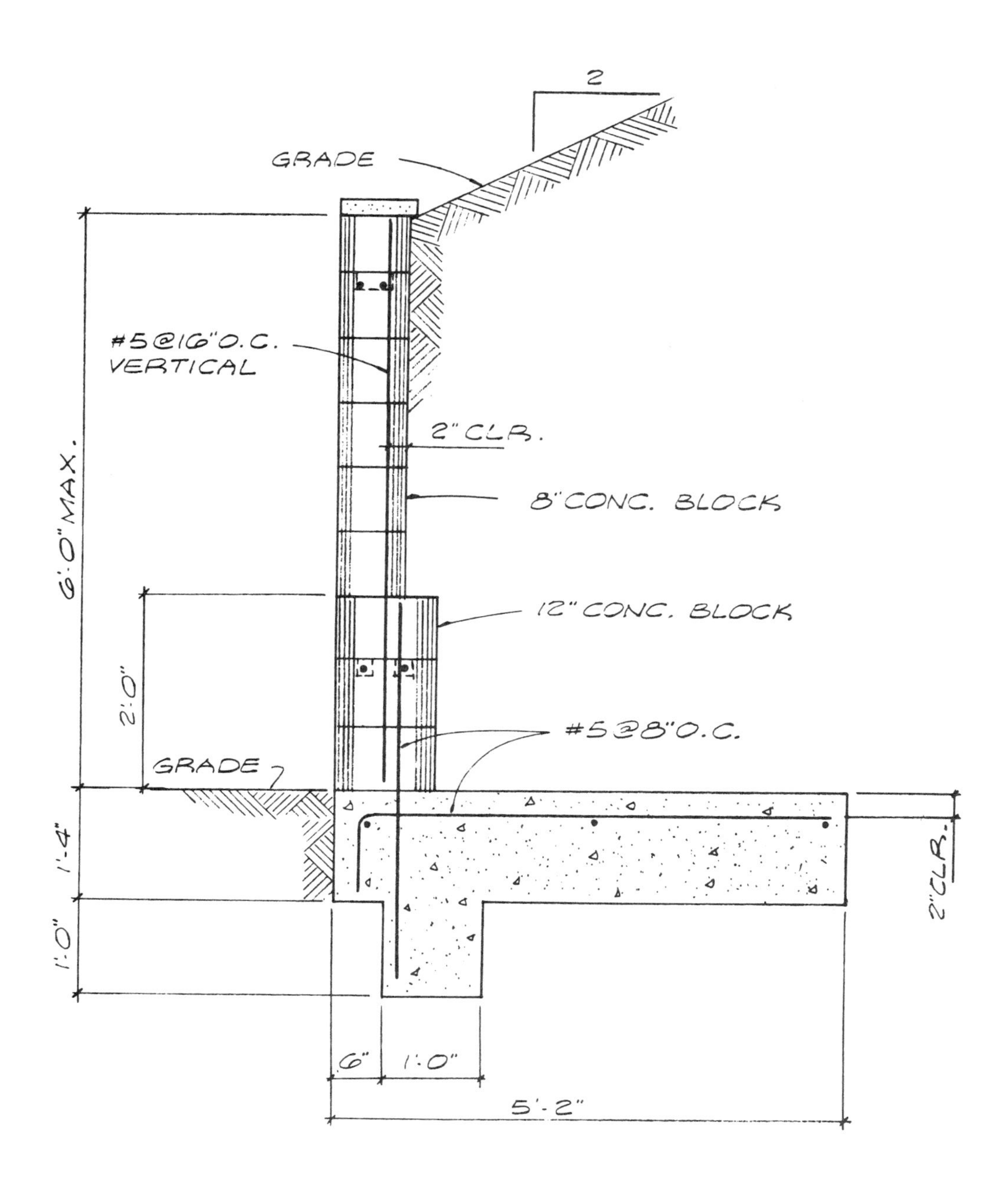

Detail 60(c). Stem height 6′0″.

Notes ▪ Drawings ▪ Ideas

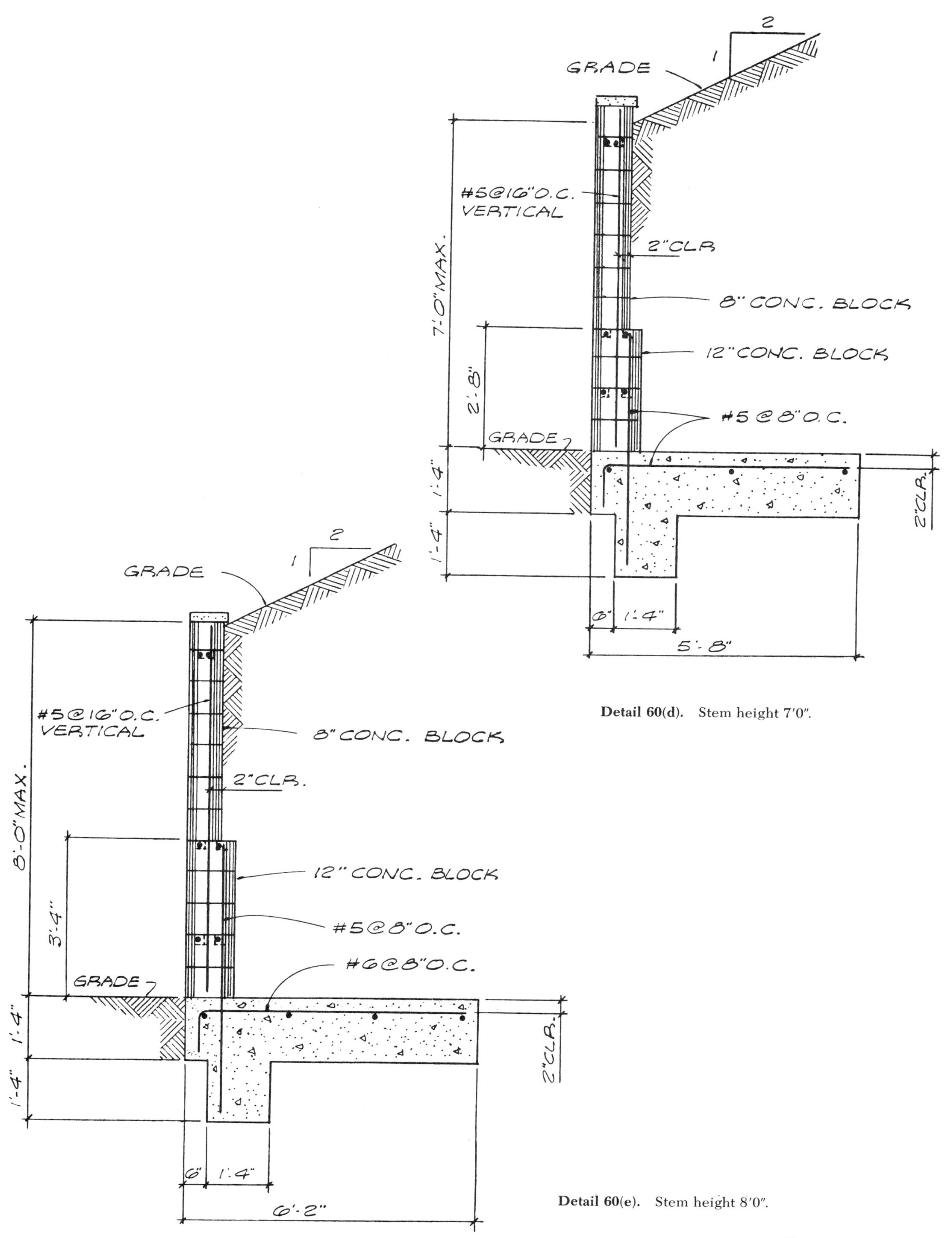

Detail 60(d). Stem height 7′0″.

Detail 60(e). Stem height 8′0″.

Notes ▪ Drawings ▪ Ideas

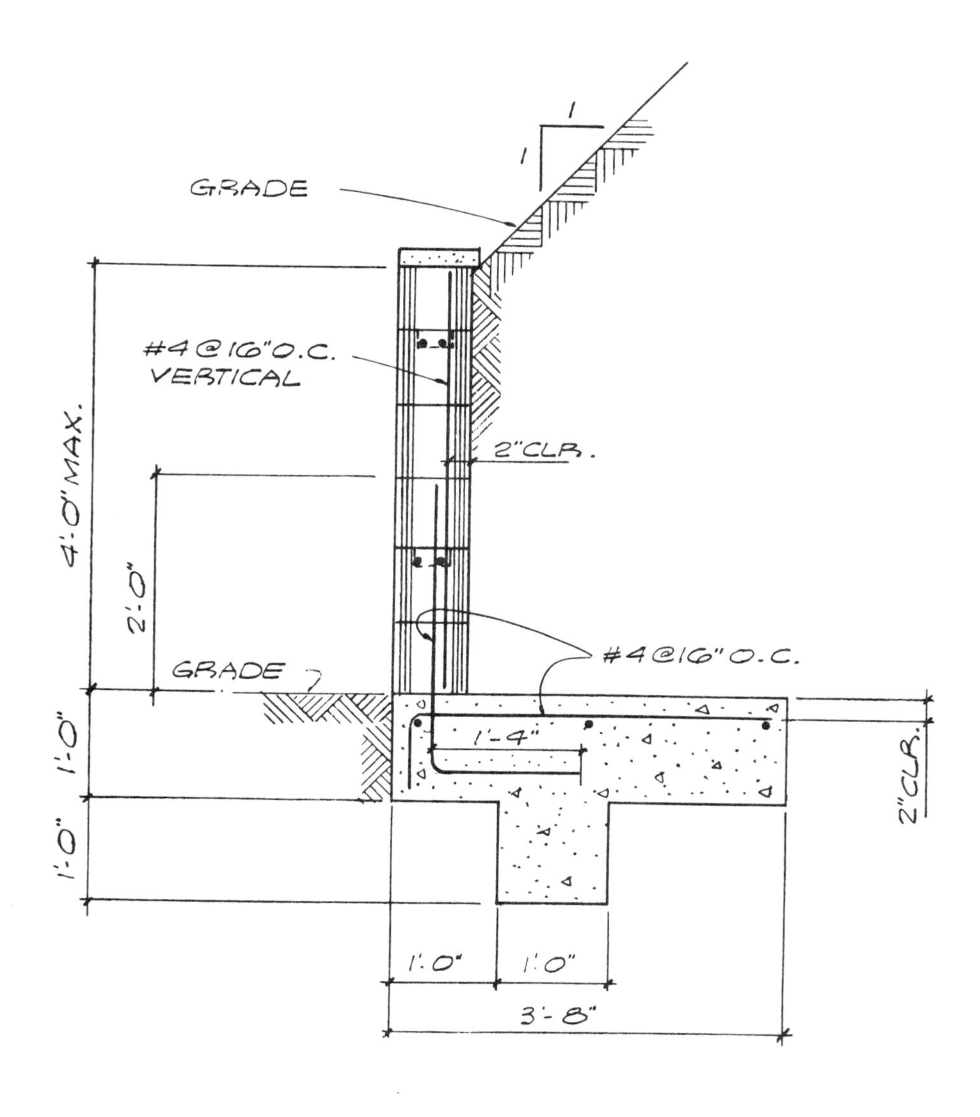

Detail 61(a). Stem height 4'0".

Detail 61. Sections of concrete block masonry retaining walls with soil at the interior face of the stem. The wall retains a sloping grade with one horizontal to one vertical and no surcharge. The wall dimensions and reinforcing steel size and spacing are determined by calculation. All concrete block masonry cells are filled with grout. Two #4 horizontal continuous reinforcing bars are placed at the top and the bottom of the wall stem. The minimum horizontal reinforcing in the stem wall and the footing is #4 at 24" o.c. The masonry head joint is omitted at 32" o.c. at the first course as a weep hole. The cement cap at the top of the wall is optional. The walls are designed for 80 lb per cu ft equivalent fluid pressure and a maximum soil bearing pressure of 1500 lb per sq. ft. The resultant of the forces passes through the middle ⅓ of the wall footing. The walls are designed to resist sliding and overturning. The overturning safety factor is 1.5. The concrete design stress is $f'c = 2000$ psi. The masonry design stress is $f'm = 600$ psi.

Notes ▪ Drawings ▪ Ideas

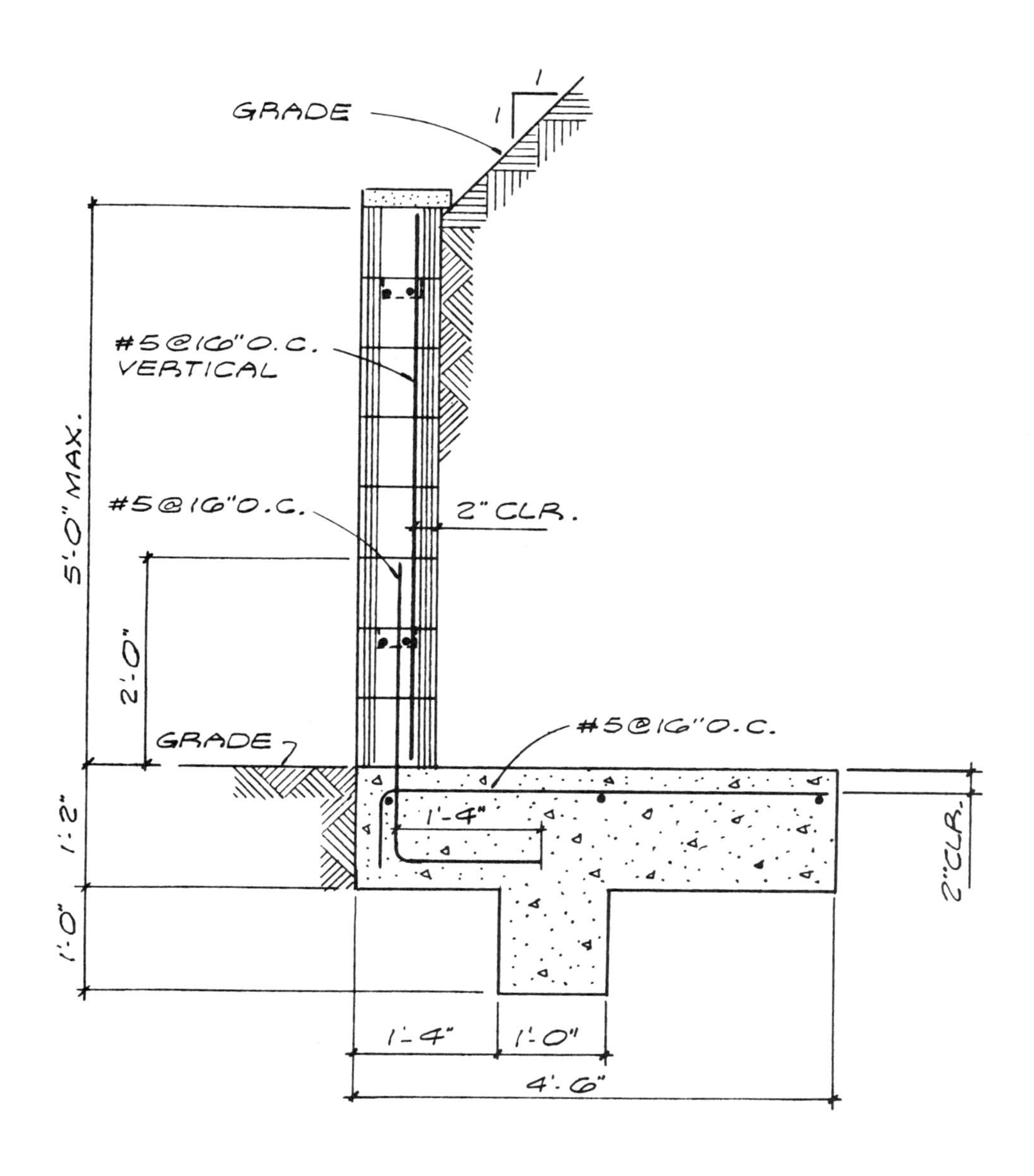

Detail 61(b). Stem height 5′0″.

Notes ▪ Drawings ▪ Ideas

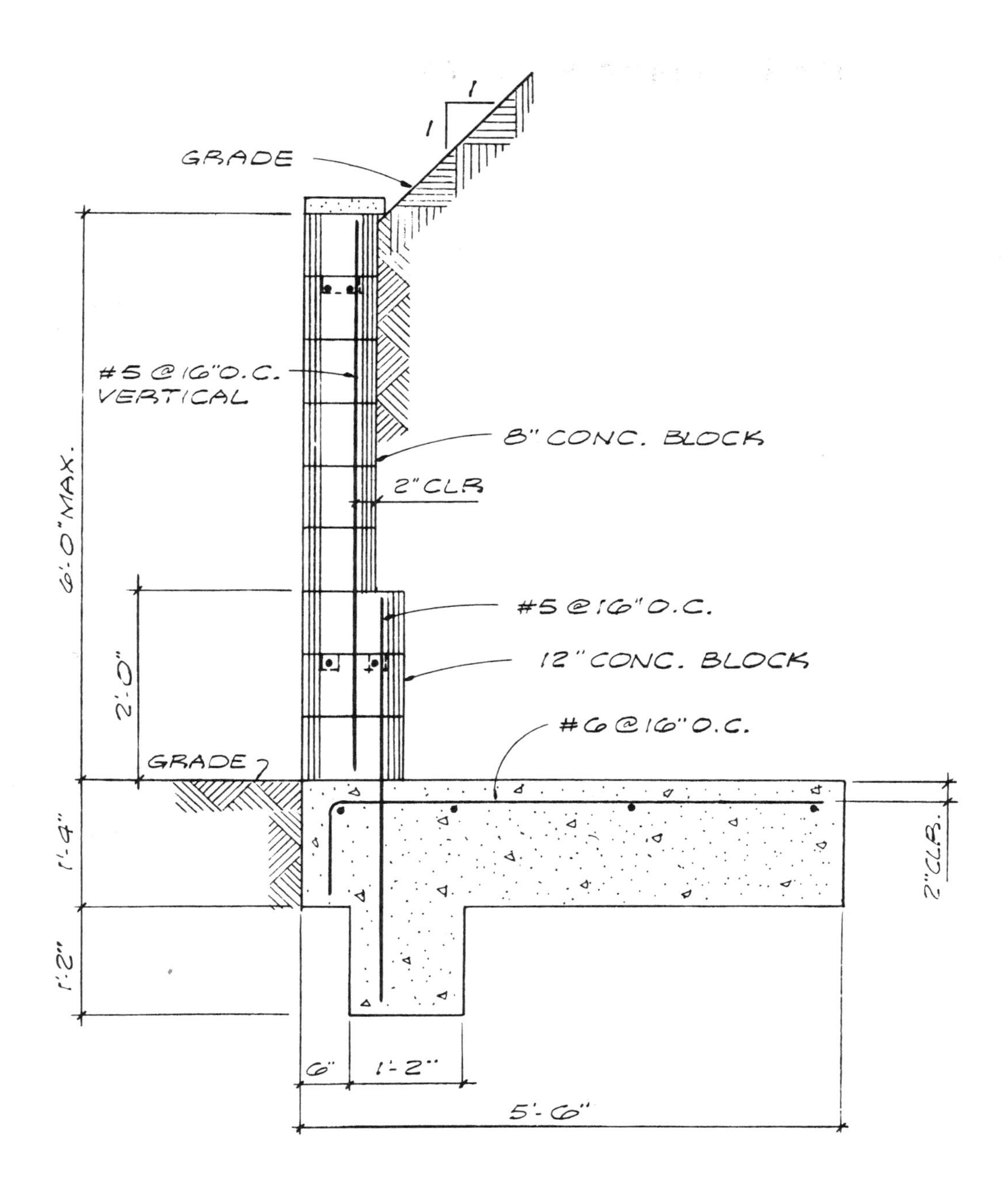

Detail 61(c). Stem height 6'0".

Notes ▪ Drawings ▪ Ideas

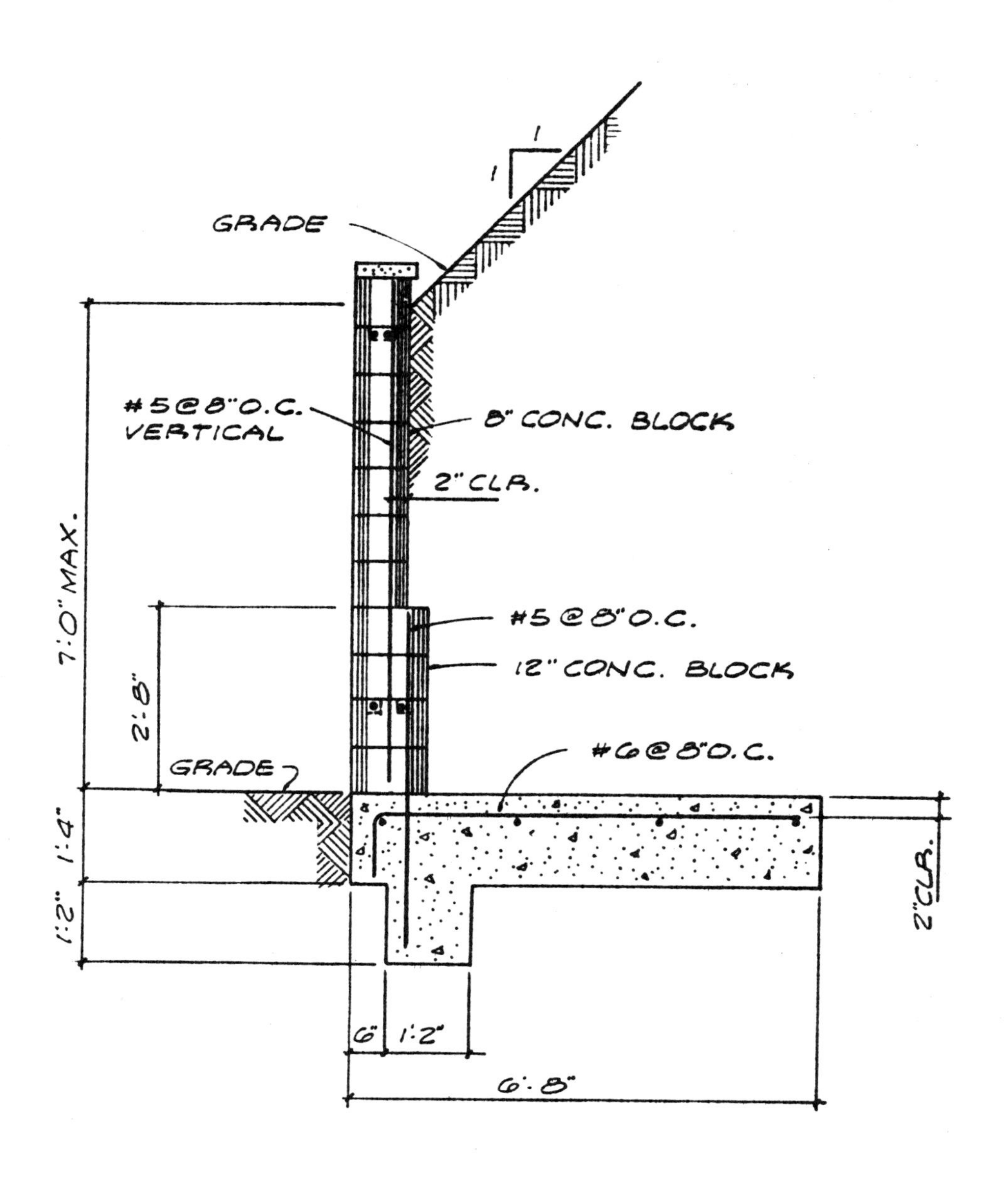

Detail 61(d). Stem height 7′0″.

Notes ▪ Drawings ▪ Ideas

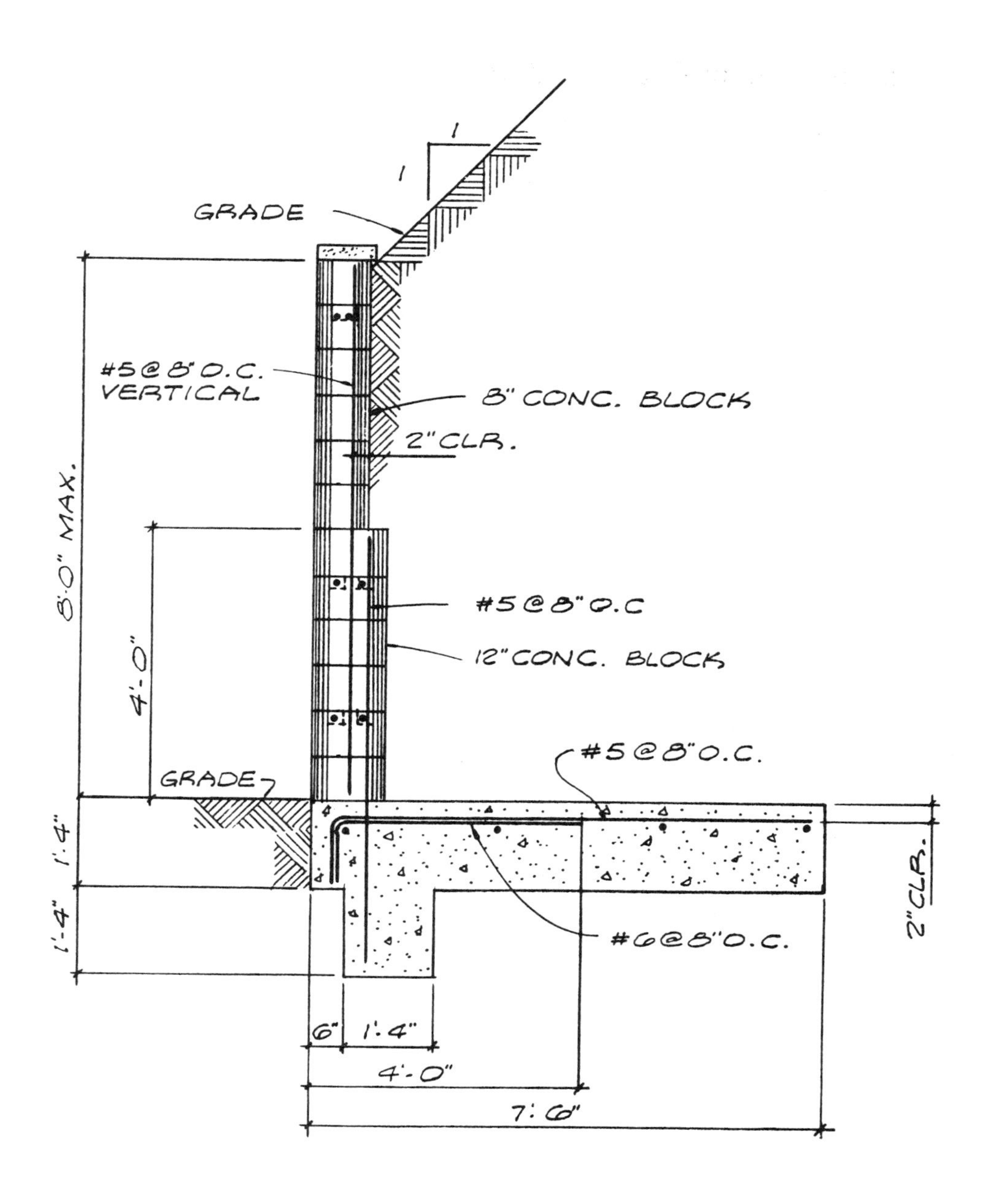

Detail 61(e). Stem height 8′0″.

Notes ▪ Drawings ▪ Ideas

INDEX